Sven Bodo Wirsing

On unit groups of modular group algebras

The concept of end-commutable ordering – with 241 exercises

Anchor Academic Publishing

Wirsing, Sven Bodo: On unit groups of modular group algebras.
The concept of end-commutable ordering – with 241 exercises,
Hamburg, Anchor Academic Publishing 2019

Buch-ISBN: 978-3-96067-224-1
PDF-eBook-ISBN: 978-3-96067-724-6
Druck/Herstellung: Anchor Academic Publishing, Hamburg, 2019

Bibliografische Information der Deutschen Nationalbibliothek:
Die Deutsche Nationalbibliothek verzeichnet diese Publikation in der Deutschen
Nationalbibliografie; detaillierte bibliografische Daten sind im Internet über
http://dnb.d-nb.de abrufbar.

Bibliographical Information of the German National Library:
The German National Library lists this publication in the German National Bibliography.
Detailed bibliographic data can be found at: http://dnb.d-nb.de

All rights reserved. This publication may not be reproduced, stored in a retrieval system
or transmitted, in any form or by any means, electronic, mechanical, photocopying,
recording or otherwise, without the prior permission of the publishers.

Das Werk einschließlich aller seiner Teile ist urheberrechtlich geschützt. Jede Verwertung
außerhalb der Grenzen des Urheberrechtsgesetzes ist ohne Zustimmung des Verlages
unzulässig und strafbar. Dies gilt insbesondere für Vervielfältigungen, Übersetzungen,
Mikroverfilmungen und die Einspeicherung und Bearbeitung in elektronischen Systemen.

Die Wiedergabe von Gebrauchsnamen, Handelsnamen, Warenbezeichnungen usw. in
diesem Werk berechtigt auch ohne besondere Kennzeichnung nicht zu der Annahme,
dass solche Namen im Sinne der Warenzeichen- und Markenschutz-Gesetzgebung als frei
zu betrachten wären und daher von jedermann benutzt werden dürften.

Die Informationen in diesem Werk wurden mit Sorgfalt erarbeitet. Dennoch können
Fehler nicht vollständig ausgeschlossen werden und die Bedey Media GmbH, die Autoren
oder Übersetzer übernehmen keine juristische Verantwortung oder irgendeine Haftung
für evtl. verbliebene fehlerhafte Angaben und deren Folgen.

Alle Rechte vorbehalten

© Anchor Academic Publishing, Imprint der Bedey Media GmbH
Hermannstal 119k, 22119 Hamburg
https://www.anchor-publishing.com, Hamburg 2019
Printed in Germany

Contents

1 Cores and normalizers — 11
  1.1 A first reduction — 11
  1.2 Cores — 20
  1.3 Normalizers — 23
  1.4 Open topics and exercises — 31

2 End-commutable orderings and exponents — 37
  2.1 Basic properties — 37
  2.2 End-commutable orderings of conjugacy classes — 44
  2.3 End-commutable orderings and nilpotent finite groups — 47
  2.4 The exponent of the center — 52
  2.5 Bounds — 57
  2.6 Open topics and exercises — 62

3 The exponent of the center for special classes of groups — 67
  3.1 The maximal possible exponent — 67
  3.2 Elementary-Abelian centers — 70
  3.3 The minimal lower bound — 73
  3.4 Central products — 76
  3.5 Wreath products — 80
  3.6 Special group extensions — 89
  3.7 Remarks to the bounds within section 2.5 — 93
  3.8 Open topics and exercises — 95

4 The invariants of the center — 101
  4.1 A direct decomposition — 101
  4.2 Commutative group algebras — 104
  4.3 The invariants — 109
  4.4 The class-graph — 113
  4.5 Determination of invariants for special cases — 115
  4.6 Isoclinism — 118
  4.7 Open topics and exercises — 122

5 Consequences for special types of unit groups 133
    5.1   Unit groups with cyclic derived subgroup 133
    5.2   Unit groups with cyclic p-power subgroup 136
    5.3   Unit groups for extra-special 2-groups 144
    5.4   Open topics and exercises 151

6 A chain of p-groups 159
    6.1   Basic properties 159
    6.2   Bounded exponents of the centers 163
    6.3   Unbounded exponents 171
    6.4   Open topics and exercises 176

List of figures 185

Bibliography 187

Index 192

For Pia, Jan and Emily

Prime factors

My heart, it holds a secret prime,
and your heart holds one, too.
I ask you now, will you join hearts,
and multiply them through?
We will make a product so sublime
it overtakes the skies
and stay together all our days
till death us factorize.

(unknown author, see [81])

The theory of groups is a central discipline within the algebra. Not only specific group theoretical methods but also concepts of other disciplines are used to analyze groups. The representation and the character theory of finite groups are two prominent examples of these methods. The exact study of the group algebra is an effectual source of insights for modules and characters. Therefor, the analysis of the group algebra has a long tradition within the theory of associative algebras (S. Jennings [30], 1941 and D.S. Passman [51], 1977). Whether the group algebra over a field K is semisimple or modular is – based on the theorem of Maschke – identifiable at the characteristic of K. Within this work the group of units of the group algebra is analyzed for a p-group and a field of characteristic p.

The structure of the group of units of the group algebra for an Abelian p-group and a finite field of characteristic p is analyzed by R. Sandling in [57], by A. Albrecht in [1] and by A. Bovdi and A. Szakacs in [13]. One main topic of this work is to determine the structure of the center of the group of units $E(KG) = (1_G + rad(KG)) \times (K \setminus \{0_K\}) \cdot 1_G$ of the group algebra KG for an non-Abelian p-group G and a field K of characteristic p.

We generalize a result of K.R. Pearson [53] at the beginning of the first chapter: for an arbitrary subgroup U of G the set $Z(G) \cap U$ is the core of U in $1_G + rad(KG)$ (corollary 1.2.3). The normalizer of U in $1_G + rad(KG)$ is determined by $N_G(U) \cdot C_{1_G + rad(KG)}(U)$ which is proven afterwards (theorem 1.3.6). The special case $U = G$ is contained in the article of D.B. Coleman (see [19]). His concept of fixed points is analyzed and generalized.

Our concept of analyzing the center of $E(KG)$ is developed within this work and is called end-commutable ordering of algebra-elements. This method is presented within the second chapter of this work. We prove (see theorem 2.3.6) that every finite group G is nilpotent if and only if every conjugacy class of G is end-commutable. In addition, we can obtain within theorem 2.1.5 for end-commutable K-algebra elements $a_1, \ldots, a_n$ the important identity $(\sum_{i=1}^{n} a_i)^{p^r} = \sum_{i=1}^{n} a_i^{p^r}$ ($p = char(K), r \in \mathbb{N}$). In the – also for our analysis – important case that $\{a_1, \ldots, a_n\}$ is a conjugacy class of a p-group also A.A. Bovdi and Z. Patay have proven this statement in [9] based on a different argumentation. We apply our method for determining the exponent of $Z(1_G + rad(KG))$ – which was also done by the same authors – purely based on characteristics of the underlying group G (theorem 2.4.8). We finalize chapter 2 by presenting some bounds for this exponent in preparation of chapter 3.

The value $\frac{|G|}{p^2}$ is the maximal possible one for the exponent of $Z(1_G + rad(KG))$ in the case of a non-Abelian p-group G (corollary 2.5.3). Within section 1 of chapter 3 we determine those groups for which this value occur: either the center of G is cyclic of order $\frac{|G|}{p^2}$ or G possesses a cyclic maximal subgroup (theorem 3.1.6).

Groups for which the center of $1_G + rad(KG)$ is elementary-Abelian are characterized in section 2 of chapter 3: the center of G is elementary-Abelian and for all $g \in G \setminus Z(G)$ the identity $C_G(g) < C_G(g^p)$ is valid (theorem 3.2.1). For example, the p-Sylow subgroups of $GL(n, GF(p^k))$ are of this kind (corollary 3.2.2.6).

In diverse interesting cases the exponent of $Z(1_G + rad(KG))$ is identical to the one of $Z(G)$: we prove this for p-groups G for which the identity $exp(G/Z(G)) \leq exp(Z(G))$ is valid (theorem 3.3.1) and – by using a complete different argument – for regular p-groups (corollary 3.3.3).

In the following sections of this chapter we analyze the exponent for group constructions. For central products of two p-groups G, H we obtain the same exponent as for their direct product which is $max\{exp(Z(1_G + rad(KG))), exp(Z(1_G + rad(KH)))\}$ (theorem 3.4.7).

We proceed the analysis by determining the exponent for an arbitrary wreath product $G \wr_\delta H$: the exponent can be calculated based on the ingredients G, H and δ (theorem 3.5.11). As a consequence, we can bound the exponent for an arbitrary action δ by the lower bound $exp(Z(1_G + rad(KG)))$ and by the upper bound $exp(Z(1_{G \times H} + rad(G \times H)))$ (corollary 3.5.18). The lower value is valid for a faithful (corollary 3.5.16) and the upper bound for the trivial action (remark 3.5.17).

For dihedral, semi-dihedral and quaternion groups of the same order the exponent of the center of $1_G + rad(KG)$ is identical. We generalize this results to two group extensions by Abelian p-groups with equivalent action and a special additional characteristic (theorem 3.6.6).

The concept of end-commutable orderings allows us not only to determine the exponent of $Z(1_G + rad(KG))$ but also the description of the p-power structure of $Z(1_G + rad(KG))$ and henceforth – for a finite field – to calculate the invariants of this Abelian p-group. We can reduce the problem to the direct factor $1_G + rad(KZ(G))$ and its co-factor $1_G + \langle \{ \sum_{x \in g^G} x \mid g \in G \backslash Z(G) \} \rangle_K$ of the center of $1_G + rad(KG)$ related to the conjugacy class sums (corollary 4.1.5). The invariants of the first factor are – as already mentioned for Abelian group algebras – completely known, and the ones of the second factor are described in two different ways (by using the chain of Frattini subgroups and the chain of socles) only by using the field K and the group G (theorem 4.3.1.3, theorem 4.3.2.6). Another description is included in the analysis of A. Bovdi and Z. Patay in [10]. The determination of the invariants and the exponent is described by using a special graph: the classgraph. The class-graph visualizes the p-power structure of the co-factor $1_G + \langle \{ \sum_{x \in g^G} x \mid g \in G \setminus Z(G) \} \rangle_K$. The exponent of the co-factor is related to the longest path within this graph, the invariants can be calculated by counting the number of special paths within it. Groups with isomorphic class-graphs possess isomorphic co-factors. We determine the invariants for some examples: we prove that the centers of $1_G + rad(KG)$ for quaternion, dihedral and semi-dihedral groups of the same order over a finite field of characteristic 2 are isomorphic (example 4.5.2.2). The last section is dedicated to isoclinic groups. We prove that the exponents of the co-factors of the center of the radicals of two isoclinic groups are identical. We use the result to describe the structure of the center of the radical for semi extra-special groups, ultra-special groups, VZ-groups, Camina and generalized Camina groups.

In chapter 5 of this work we prove at first that the derived subgroup of $1_G + rad(KG)$ is cyclic only for Abelian G (theorem 5.1.4). Afterwards we prove that $(1_G + rad(KG))^p$ is cyclic if and only if G is elementary-Abelian or G is Abelian and $p = |G^2| = |K^2| = 2$ is valid (corollary 5.2.11).
The group $1_G + rad(KG)$ is special only for an extra-special 2-group G (proposition 5.3.9). For such a group G the elementary-Abelian center of $1_G + rad(KG)$ is identical to the Frattini subgroup of $1_G + rad(KG)$ and contains all squares (lemma 5.3.2, theorem 5.3.3). Within this work we do not give a description of all groups G such that $1_G + rad(KG)$ is a special 2-group. But in the smallest relevant case we prove that the derived subgroup of $1_G + rad(KG)$ is of index 2 in $Z(1_G + rad(KG))$ (example 5.3.10).

Within chapter 6 we focus on the chain of iterated p-groups defined by $G_0 := G$ and $G_{n+1} := 1 + rad(KG_n)$ for all $n \in \mathbb{N}$ over a finite field K of characteristic p and a non-Abelian p-group G. The previous chapters are linked to $G_1 = 1 + rad(KG)$. Now we want to study the behavior of this chain. Several parameters of this chain turn out to be increasing resp. unbounded (see proposition 6.1.2, e.g. the corresponding chain of derived subgroups, of breadth, of nilpotency classes, of strong derived length, of class numbers, of Baer-length). As a consequence the corresponding chain of degrees of commutativities converges against zero.

But the structure of the centers related to this chain can be described differently: the exponents are stable after the second step because the direct factor related to the class sums is elementary-p-Abelian. This result is generalized to arbitrary radical algebras (see theorem 6.2.11 and corollary 6.2.12). As a consequence we can prove that the chain of corresponding exponents and the chain of Engel-length of $(G_n)_{n \in \mathbb{N}_0}$ are unbounded (see theorem 6.3.3).

Some applications are also transferred to the exercises at the end of each chapter. Some exercises are included enhancing the theory presented so far. In addition, at the beginning of each exercise series some open-ended topics are included which can be used by the reader – and also by the author – to do additional researches within this theory. The author has included some graphics – mostly so called Hasse diagrams – to visualize the main results of this work.

The author has prepared some slides which can be used as a basic for a presentation of this work. These slides are available and can be requested at the Anchor Academic Publishing service by using the email address info@anchor-publishing.com.

List of symbols

In this chapter we list all symbols used in this work, present a short description and link (section and page) the first appearance of the symbol within this work.

Chapter 1

$*$	the star composition; 1.1.1, 11
$cl(A)$	class of nilpotency of an associative algebra; exercise 27, 33
$Q(A)$, A^*	the group of units of the monoid $(A; *)$; 1.1.3, 11
a'	the inverse of $a \in Q(A)$; 1.1.3, 11
$E(A)$	the group of units of an associative unitary algebra A; 1.1.3, 11
(I, T)	a semidirect decomposition of an algebra; 1.1.6, 12
(N, U)	a semidirect decomposition of a group; 1.1.6, 12
$S + T$	$:= \{s + t \mid (s; t) \in S \times T\}$; 1.1.8, 13
$s + T$	$:= \{s\} + T$; 1.1.8, 13
$\overline{M}$	$:= \sum_{m \in M} m$; 1.1.9, 13
n_K	$:= \sum_{i=1}^{n} 1_K$; 1.1.9, 13
$\lvert T \rvert$	the order of a finite set T; 1.1.9, 13
e_H	$:= \frac{1}{\lvert H \rvert} \overline{H}$; 1.1.9, 13
KM	the free K-module with K-basis M ; 1.1.9, 13
$Z(A)$	the center of an algebra A; 1.1.10, 13
$\cong_K$	the isomorphism within the class of K-spaces; 1.1.11, 14
$\langle \ldots \rangle_K$	the K-span within a K-space; 1.1.11, 14
$\mathcal{A}$	the class of associative algebras; 1.1.11, 14
$\mathcal{A}_1$	the class of associative unitary algebras; 1.1.11, 14
$\mathcal{L}$	the class of Lie algebras; 1.1.11, 14
$\mathcal{G}$	the class of groups; 1.1.11, 14
$\cong_{\mathcal{K}}$	the isomorphism within the class $\mathcal{K}$; 1.1.11, 14
$\langle \ldots \rangle_{\mathcal{K}}$	the span within the class $\mathcal{K}$; 1.1.11, 14
$\mathbb{N}$	the set of natural numbers; 1.1.11, 14
$\mathbb{H}$	the set of real quaternions; exercise 9, 31
$\underline{n}$	$:= \mathbb{N}_{\leq n}$; 1.1.11, 14

7

$\underline{n}_0$	$:= \underline{n} \cup \{0\}$; 1.1.11, 14
$Aug_B(V)$	$:= \langle \{b_1 - b_2 \mid b_1, b_2 \in B\} \rangle_K$; 1.1.12, 14
$aug_B(\sum_{b \in B} k_b b)$	$:= \sum_{b \in B} k_b$; 1.1.12, 14
$Aug(KM)$	$:= Aug_M(KM)$; 1.1.12, 14
$aug(x)$	$:= aug_M(x)$, $x \in KM$; 1.1.12, 14
aug	the augmentation function; 1.1.13, 14
$\ker \alpha$	the kernel of the function α; 1.1.13, 14
G/N	the factor group of G modulo N; 1.1.14, 14
Ng	an element of G/N; 1.1.14, 14
p_N	the linearization of $g \mapsto Ng$; 1.1.14, 14
$T \cdot S$	$:= \langle \{ts \mid (t,s) \in T \times S\} \rangle_K$; 1.1.14, 14
$rad(A)$	the nilradical of an associative algebra A; 1.1.15, 17
$o(g)$	the order of an element g of a group; 1.1.15, 17
$Z(G)$	the center of the group G; 1.1.15, 17
$char(K)$	the characteristic of the field K; 1.1.15, 17
Q_n	the quaternion group of order n; 1.1.19, 19
$core_G(U)$	the core of U in G; 1.2.1, 20
g^h	$:= h^{-1}gh$; 1.2.2, 20
T^h	$:= \{t^h \mid t \in T\}$; 1.2.2, 20
$Abb(M,N)$	the set of functions between M and N; 1.3.1, 23
$\bar{\delta}$	the linearization of δ; 1.3.3, 23
$\hat{\delta}$	enhanced group action with respect to δ; 1.3.3, 23
$\alpha_{|T}$	the restriction of α to T; 1.3.2, 23
$N_G(U)$	the normalizer of U in G; 1.3.6, 24
$C_G(U)$	the centralizer of U in G; 1.3.6, 24
$[g,h]$	the commutator of g and h; 1.3.10, 27
$c(G)$	the class number of G; 1.3.10, 27
$U \oplus_K W$	the inner direct sum of the K-subspaces U and W; 1.3.11, 27
$dim_K(V)$	the dimension of the K-space V; 1.3.11, 27
$C_{KM,\delta}(U)$	the centralizer of U in KM with respect to δ; 1.3.13, 29
κ_g	the conjugation with g; 1.3.15, 29
κ	the function $g \mapsto \kappa_g$; 1.3.15, 29
g^G	the conjugacy class of g in G; 1.3.11, 27

Chapter 2

$EA(T)$	the set of end-commutable orderings of a set T; 2.1.1, 37
S_n	the symmetric group on $\underline{n}$; 2.1.2, 37
D_n	the dihedral group of order n; 2.1.2, 37
$C_A(T)$	the centralizer of T in A; 2.1.5, 38
$Aut(G)$	the group of automorphism of G; 2.1.8, 40
$Stab_G(m)$	the stabilizer of m in G; 2.1.8, 40
$Inn(G)$	the group of inner automorphism of G; 2.1.9, 41

V_4	the Klein four-group; 2.1.9, 41
G'	the derived subgroup of G; 2.2.1, 44
$\Phi(G)$	the Frattini subgroup of G; 2.2.1, 44
$F(G)$	the Fitting subgroup of G; 2.3.4, 48
$\mathbb{C}$	the complex number field; 2.3.8, 50
φ_T	the monotone bijection between $\mid T \mid$ and T; 2.4.1, 52
$\binom{T}{i}$	the set of subsets of order i of T; 2.4.2, 52
$\mathbb{N}_0$	$:= \mathbb{N} \cup \{0\}$; 2.4.2, 52
$\binom{n}{i}$	$:= \mid \binom{n}{i} \mid$; 2.4.4, 53
G^n	$:= \langle \{g^n \mid g \in G\} \rangle_{\mathcal{G}}$; 2.4.5, 53
K^{p^n}	$:= \{k^{p^n} \mid k \in K\}$, K a field; 2.4.5, 53
$exp(G)$	the exponent of a torsion group G; 2.4.5, 53
$max\, T$	the maximum of a finite subset T of $\mathbb{N}$; 2.4.7, 54
$min\, T$	the minimum of a finite subset T of $\mathbb{N}$; 2.4.8, 55
$C_G(g)$	$:= C_G(\{g\})$; 2.4.8, 55
$a \circ b$	$:= ab - ba$; 2.4.9, 56
A°	the associated Lie algebra of A; 2.4.9, 56
$\mathcal{K}(G)$	the set of conjugacy classes of G; 2.5.1, 57
$gcd(a,b)$	greatest common divisor of a,b; exercise 46, 63

Chapter 3

SD_n	the semi dihedral group of order n; 3.1.2, 67
Z_n	the cyclic group of order n; 3.1.6, 69
$GL(n,K)$	the general linear group; 3.2.2.1, 71
$GF(p^k)$	the finite field with p^k elements; 3.2.2.1, 71
P_n	a p-Sylow subgroup of $GL(n, GF(p^k))$; 3.2.2.1, 71
$K^{n \times n}$	$:= K^{\underline{n} \times \underline{n}}$; 3.2.2.1, 71
$E_{i,j}$	a basis vector of $K^{n \times n}$; 3.2.2.2, 71
$su(n,K)$	strict lower triangular matrices of $K^{n \times n}$; 3.2.2.2, 71
$PGL(n,K)$	the projective linear group; 3.2.2.6, 72
$SL(n,K)$	the special linear group; 3.2.2.6, 72
$PSL(n,K)$	the projective special linear group; 3.2.2.6, 72
$G \times H$	the direct product of two groups G, H; 3.4.1, 76
D_μ, D	$:= \{(u; (u\mu)^{-1}) \mid u \in U_1\}$; 3.4.1, 76
$G_1 \curlyvee_\mu G_2, G_1 \curlyvee G_2$	the central product of two groups G, H; 3.4.1, 76
A^B	$:= Abb(B,A)$; 3.4.9, 79
$a \equiv b\, mod\, c$	c divides $a - b$; 3.4.11, 79
$\varphi^s, \tilde{s}$	two special functions; 3.5.1, 81
$H \wr_\delta S, H \wr_X S$	the wreath product of two groups G, H; 3.5.2, 81
$H \wr S$	the regular wreath product of H with S; 3.5.2, 81
$\alpha \equiv h$	the constant function with value h; 3.5.4, 81
$G/_r U$	the set of right cosets of U in G; 3.5.6, 82
$[A,B]$	$:= \langle \{[a,b] \mid a \in A, b \in B\} \rangle_{\mathcal{G}}$; 3.5.7, 83

$Fix_X(g)$	$:= \{x \mid x \in X, xg = x\}$; 3.5.7,	83
α_h	the constant function with unique value h; 3.5.9,	84
$C(2n, q)$	the symplectic group; 3.5.15,	87
$U(n, q^2)$	the unitary group; 3.5.15,	87
$O_D(n, q)$	the orthogonal group; 3.5.15,	87
$(H \times N; \cdot_{\alpha, N(\cdot;\cdot)})$	the group extension of G by N; 3.6.2,	90
$N_R(\cdot;\cdot)$	the factor system for the representive system R; 3.6.1,	89
$\alpha_R(h)$	the automorphism for the representive system R; 3.6.1,	89

Chapter 4

$\overline{\mathcal{K}(G)}$	$:= \langle \{\overline{g^G} \mid g \in G \setminus Z(G)\} \rangle_K$; 4.1.6,	103
nG	another symbol for $G^{\underline{n}}$; 4.2.1.1,	104
$\overline{k(G)}_{p^i}$	the dimension of $(\overline{\mathcal{K}(G)}^*)^{p^i}$; 4.3.1.2,	109
$soc_n(G)$	the n-th socle of G; 4.3.2.1,	110
$\sim_n$	a special equivalence relation on $\mathcal{K}(G) \setminus \{\{z\} \mid z \in Z(G)\}$; 4.3.2.4,	111
$\log$	logarithm; exercise 135,	123
$[,]_G$	the commutator map of a group G; 4.6.2,	119

Chapter 5

A^n	$:= \langle \{a_1 \ldots a_n \mid a_i \in A\} \rangle_K$; 5.1.2,	133
$a_1 \circ \cdots \circ a_n$	$:= (\ldots (a_1 \circ a_2) \circ \ldots) \circ a_n$; 5.2.1,	137
$cl(G)$	the class of nilpotency for a nilpotent group G; 5.2.2,	137
$cl(L)$	the class of nilpotency for a nilpotent Lie algebra L; 5.2.2,	137
$Z_n(G)$	the n-th center of a group G; 5.2.5, before	139
$Z_n(L)$	the n-th center of a Lie algebra L; before 5.2.5, before	139
$L \circ L$	$:= \langle \{a \circ b \mid a, b \in L\} \rangle_K$; 5.2.6,	140
U_{even}	a special subgroup of $E(KG)$; 5.3.7,	147

Chapter 6

$J(A)$	Jacobson radical of an algebra A; 6.2.5,	165
$L^{(n)}$	n-th term of the lower central chain of a Lie algebra L; 6.2.5,	165
$a^{(b)}$	star conjugate of a with b; 6.2.5,	165
$ad(a)$	right multiplication with a in a Lie algebra; 6.2.10,	168
$b(G)$	breadth of a group G; 6.1.2, before	160
$st(A)$	solvable class or derived length of an algebra A; 6.1.2, before	160
$(G_n)_{n \in \mathbb{N}_0}$	a special chain of p-groups; 6.1.1,	159
W_n	n-th term of a special upper central chain; 6.2.5,	165
X_n	n-th term of a special upper central chain; 6.2.5,	165
$\gamma_n(G)$	n-th term of the lower central chain of G; 6.2.5,	165
$d(G)$	degree of commutativity of G; 6.1.2,	160

Chapter 1

Cores and normalizers

1.1 A first reduction

Within this work a K-algebra is an algebra defined based on a commutative unitary ring K.

1.1.1 Definition (star composition)
Let A be a K-algebra. For all $a, b \in A$ we define
$$a * b := a + b + ab.$$
B.L. van der Waerden calls $*$ the star composition on A.⋄

1.1.2 Remark ($*$ versus $\cdot$)
For every associative K-algebra A the following statements are valid:

(i) $(A; *)$ is a monoid possessing the unit element 0_A.

(ii) If A is unitary, then the function $A \to A$, $a \mapsto 1_A + a$ is a monoid isomorphism between $(A; *)$ and $(A; \cdot)$.⋄

1.1.3 Definition (star group)
If A is an associative K-algebra, then we denote by $Q(A)$ the group of units of the monoid $(A; *)$ and for every $a \in Q(A)$ by a' the inverse of a in $Q(A)$. The elements of $Q(A)$ are called star regular or quasi regular and the group $Q(A)$ is called the star or quasi regular group of A. If A is unitary, then $E(A)$ is called the group of units of A.⋄

The following remark shows us that the star group is a generalization of the group of units in the context of non-unitary associative algebras.

1.1.4 Remark

For every K-algebra A the following statements are valid:

(i) For all $a, b, c, d \in A$ the identity $(a+b) * (c+d) = a*c + b*d + ad + bc$ is true.

(ii) If A is associative, then for all $a, t \in Q(A)$ the identity $a' * t * a = t + a't + ta + a'ta$ is valid.

(iii) If A is associative and unitary, then the restriction of the function $A \to A$, $a \mapsto 1_A + a$ to $Q(A)$ is a group isomorphism between $Q(A)$ and $E(A)$. ⋄

1.1.5 Proposition

For every associative K-algebra A the following statements are valid:

(i) For every subalgebra T of A the set $Q(T)$ is a subgroup of $Q(A)$.

(ii) For every ideal I of A the set $Q(I) = Q(A) \cap I$ is a normal subgroup of $Q(A)$.

Proof. ad(i): This statement is straightforward to prove.

ad(ii): Because of part (i) the set $Q(I)$ is a subgroup of $Q(A)$. For all $a \in Q(A) \cap I$ the identity $a' = -a - aa' \in I$ is valid, and hence $Q(A) \cap I = Q(I)$ is deduced. If $t \in Q(I)$ and $a \in Q(A)$ are valid, then we use part (ii) of remark 1.1.4 to conclude $a' * t * a = t + a't + ta + a'ta$. Thus, $a' * t * a \in Q(A) \cap I = Q(I)$ is proven. ⋄

1.1.6 Definition (semidirect decomposition)

If A is a K-algebra, then we call a pair (I, T) a semidirect resp. direct decomposition of A, if A is the inner direct sum of the ideal I and the subalgebra resp. the ideal T of A.

For a group G a pair (N, U) is called a semidirect resp. direct decomposition of G, if G is the product of the normal subgroup N and of the subgroup resp. the normal subgroup U of G and $N \cap U = \{1_G\}$ is valid. ⋄

1.1.7 Proposition

If A is an associative K-algebra and (I, T) a semidirect decomposition of A, then $(Q(I), Q(T))$ is a semidirect decomposition of $Q(A)$.

Proof. By using proposition 1.1.5 the set $Q(I)$ is a normal subgroup and $Q(T)$ is a subgroup of $Q(A)$ such that their intersection is exactly $\{0_A\}$.

Cores and normalizers   13

Let $q \in Q(A)$. Elements $i, j \in I$ and $t, s \in T$ exist such that $q = i + t$ and $q' = j + s$ are valid. Because of $0_A = q * q'$ and part (i) of remark 1.1.4 we deduce $0_A = t * s + i * j + tj + is$, and thus $t * s = 0_A$ is valid. The identity $0_A = q' * q$ is used to prove $s * t = 0_A$ in a similar way. Hence, $t \in Q(T)$ and $t' = s$ are true. We use part (i) of remark 1.1.4 to conduct $(i+is) * t = i*t + is + ist = i+t+it+is+ist = i+t+i(s*t) = q$. Because of $q \in Q(A)$ and $t \in Q(T)$ we deduce $i + is \in Q(A) \cap I$, and by using part (ii) of proposition 1.1.5 the proof is finished.⋄

1.1.8 Corollary

Let (I, T) be a semidirect decomposition of an associative K-algebra A. The following statements are valid:

(i) If T is an ideal of A or T is central in A, then $(Q(I), Q(T))$ is a direct decomposition of $Q(A)$.

(ii) If A is unitary, then $(1_A + Q(I), 1_A + Q(T))$ is a semidirect decomposition of $E(A)$.

(iii) If A is unitary and $(Q(I), Q(T))$ is a direct decomposition of $Q(A)$, then $(1_A + Q(I), 1_A + Q(T))$ is a direct decomposition of $E(A)$.

Proof. ad(i): This statement is a direct consequence of proposition 1.1.7 and part (ii) of proposition 1.1.5 because $Q(T)$ is a normal subgroup of $Q(A)$ in the mentioned scenarios.

ad(ii) and (iii): These statements are a consequence of proposition 1.1.7 and part (iii) of proposition 1.1.5.⋄

1.1.9 Definition

(i) If K is a field and $n \in \mathbb{N}_0$, then we define $n_K := \sum_{i=1}^{n} 1_K$.

(ii) For every finite subset M of a K-algebra A we define $\overline{M} := \sum_{m \in M} m$. For a group G, a finite and non-empty subset H of G and a field K such that $char(K)$ is not dividing $\mid H \mid$ we define $e_H := \frac{1}{|H|_K} \overline{H}$.⋄

1.1.10 Proposition (idempotents and subgroups)

Let G be a group, H a finite and non-empty subset of G and K a field such that $char(K)$ is not dividing $\mid H \mid$. e_H is an idempotent of KG if and only if H is a subgroup of G.

Proof. If H is a subgroup of G, then for all $h \in H$ the identity $h\overline{H} = \overline{H}$

is valid, and thus $\overline{H}^2 = |H|_K \overline{H}$ and $(e_H)^2 = e_H$ are valid.
If e_H is an idempotent of KG, then $\overline{H}^2 = |H|_K \overline{H}$ is true. Let $x, y \in H$. Elements $k \in K$ and $h \in H$ exist such that $kxy = |H|_K h$ is valid. If $xy \neq h$ would be true, then $|H|_K = 0_K$ would be valid which is a contradiction. Therefor $xy = h \in H$ is proven, and by using the finiteness of H we deduce that H is a subgroup of G.◇

1.1.11 Definition (isomorphism and span)

(i) If K is a field, then $\cong_K$, $\langle \ldots \rangle_K$ etc. are called the isomorphism, the span etc. within the class of K-spaces.
By $\mathcal{A}, \mathcal{A}_1, \mathcal{L}$ resp. $\mathcal{G}$ we denote the class of associative algebras over K, the class of associative unitary algebras over K, the class of Lie algebras over K resp. the class of groups. If $\mathcal{X}$ is one of these classes, then we denote by $\cong_{\mathcal{X}}$, $\langle \ldots \rangle_{\mathcal{X}}$ etc. the isomorphism, the span etc. within the class $\mathcal{X}$.

(ii) For all $n \in \mathbb{N}$ we define $\underline{n} := \mathbb{N}_{\leq n}$ and $\underline{n}_0 := \underline{n} \cup \{0\}$.◇

1.1.12 Definition (augmentation)

Let K be a field and V a finite-dimensional K-space. For every K-basis B of V we define $Aug_B(V) := \langle \{b_1 - b_2 \mid b_1, b_2 \in B\} \rangle_K$. If $v \in V$, then for every $b \in B$ exactly one $k_b \in K$ exists such that $v = \sum_{b \in B} k_b b$ is valid, and we define $aug_B(v) := \sum_{b \in B} k_b$. For a finite magma M we use the notation $Aug(KM) := Aug_M(KM)$ and call $Aug(KM)$ the augmentation ideal of KM. If $x \in KM$, then $aug(x) := aug_M(x)$ is defined and called the augmentation of x.◇

1.1.13 Remark (augmentation ideal)

Let K be a field, M a finite non-empty magma and $aug : KM \to K$ the K-linear extension of the function $M \to K$, $m \mapsto 1_K$. The augmentation function aug is an algebra-epimorphism such that $ker(aug) = Aug(KM)$ is valid. In particular, $Aug(KM)$ is an ideal of codimension 1 (and hence a maximal ideal) of KM. For every $m \in M$ the set $\{x - m \mid x \in M \setminus \{m\}\}$ is a K-basis of $Aug(KM)$.◇

1.1.14 Definition and remark (kernel of the augmentation map)

Let K be a field, G a finite group, N a normal subgroup of G and $p_N : KG \to K(G/N)$ the linearization of the $\mathcal{G}$-epimorphism $G \to G/N$, $g \mapsto Ng$. By using lemma 1.8 of chapter 1 in [51] the kernel of p_N is exactly $KG\, Aug(KN) = Aug(KN)\, KG$.◇

Cores and normalizers 15

For the following lemma several proofs are existing (see e.g. D.A.R. Wallace in [74], L.E. Dickson[1] in [23] or R.L. Kruse and D.T. Price in [37]). We

---

[1]Leonard Eugene Dickson (January 22, 1874 to January 17, 1954) was an American mathematician. He was one of the first American researchers in abstract algebra, in particular the theory of finite fields and classical groups, and is also remembered for a three-volume history of number theory, History of the Theory of Numbers.

Dickson considered himself a Texan by virtue of having grown up in Cleburne, where his father was a banker, merchant, and real estate investor. He attended the University of Texas at Austin, where George Bruce Halsted encouraged his study of mathematics. Dickson earned a B.S. in 1893 and an M.S. in 1894, under Halsted's supervision. Dickson first specialised in Halsted's own specialty, geometry.

Both the University of Chicago and Harvard University welcomed Dickson as a Ph.D. student, and Dickson initially accepted Harvard offer, but chose to attend Chicago instead. In 1896, when he was only 22 years of age, he was awarded Chicago's first doctorate in mathematics, for a dissertation titled The Analytic Representation of Substitutions on a Power of a Prime Number of Letters with a Discussion of the Linear Group, supervised by E. H. Moore.

Dickson then went to Leipzig and Paris to study under Sophus Lie and Camille Jordan, respectively. On returning to the USA, he became an instructor at the University of California. In 1899 and at the extraordinarily young age of 25, Dickson was appointed associate professor at the University of Texas. Chicago countered by offering him a position in 1900, and he spent the balance of his career there. At Chicago, he supervised 53 Ph.D. theses; his most accomplished student was probably A. A. Albert. He was a visiting professor at the University of California in 1914, 1918, and 1922. In 1939, he returned to Texas to retire.

Dickson married Susan McLeod Davis in 1902; they had two children, Campbell and Eleanor.

Dickson was elected to the National Academy of Sciences in 1913, and was also a member of the American Philosophical Society, the American Academy of Arts and Sciences, the London Mathematical Society, the French Academy of Sciences and the Union of Czech Mathematicians and Physicists. Dickson was the first recipient of a prize created in 1924 by The American Association for the Advancement of Science, for his work on the arithmetics of algebras. Harvard (1936) and Princeton (1941) awarded him honorary doctorates.

Dickson presided over the American Mathematical Society in 1917 to 1918. His December 1918 presidential address, titled 'Mathematics in War Perspective,' criticized American mathematics for falling short of those of Britain, France, and Germany: 'Let it not again become possible that thousands of young men shall be so seriously handicapped in their Army and Navy work by lack of adequate preparation in mathematics.' In 1928, he was also the first recipient of the Cole Prize for algebra, awarded annually by the AMS, for his book Algebren und ihre Zahlentheorie.

It appears that Dickson was a hard man: 'A hard-bitten character, Dickson tended to speak his mind bluntly; he was always sparing in his praise for the work of others. ... he indulged his serious passions for bridge and billiards and reportedly did not like to lose at either game. He delivered terse and unpolished lectures and spoke sternly to his students. ... Given Dickson's intolerance for student weaknesses in mathematics, however, his comments could be harsh, even though not intended to be personal. He did not aim to make students feel good about themselves. Dickson had a sudden death trial for his prospective doctoral students: he assigned a preliminary problem which was shorter than a dissertation problem, and if the student could solve it in three months, Dickson would agree to oversee the graduate student's work. If not the student had to look elsewhere for an advisor.'

Dickson had a major impact on American mathematics, especially abstract algebra. His mathematical output consists of 18 books and more than 250 papers. The Collected

present another approach.

Mathematical Papers of Leonard Eugene Dickson fill six large volumes.

In 1901, Dickson published his first book Linear groups with an exposition of the Galois field theory, a revision and expansion of his Ph.D. thesis. Teubner in Leipzig published the book, as there was no well-established American scientific publisher at the time. Dickson had already published 43 research papers in the preceding five years; all but seven on finite linear groups. Parshall (1991) described the book as follows: 'Dickson presented a unified, complete, and general theory of the classical linear groups – not merely over the prime field $GF(p)$ as Jordan had done – but over the general finite field $GF(p^n)$, and he did this against the backdrop of a well-developed theory of these underlying fields. His book represented the first systematic treatment of finite fields in the mathematical literature.' An appendix in this book lists the non-Abelian simple groups then known having order less than 1 billion. He listed 53 of the 56 having order less than 1 million. The remaining three were found in 1960, 1965, and 1967. Dickson worked on finite fields and extended the theory of linear associative algebras initiated by Joseph Wedderburn and Cartan. He started the study of modular invariants of a group. In 1905, Wedderburn, then at Chicago on a Carnegie Fellowship, published a paper that included three claimed proofs of a theorem stating that all finite division algebras were commutative, now known as Wedderburn's theorem. The proofs all made clever use of the interplay between the additive group of a finite division algebra A, and the multiplicative group. Karen Parshall noted that the first of these three proofs had a gap not noticed at the time. Dickson also found a proof of this result but, believing Wedderburn's first proof to be correct, Dickson acknowledged Wedderburn's priority. But Dickson also noted that Wedderburn constructed his second and third proofs only after having seen Dickson's proof. She concluded that Dickson should be credited with the first correct proof.

Dickson's search for a counterexample to Wedderburn's theorem led him to investigate non-associative algebras, and in a series of papers he found all possible three and four-dimensional (non-associative) division algebras over a field.

In 1919 Dickson constructed Cayley numbers by a doubling process starting with quaternions. His method was extended to a doubling of the real numbers to produce the complex numbers, and of the complex numbers to produce the real quaternions by A. A. Albert in 1922, and the procedure is known now as the Cayley-Dickson construction of composition algebras.

Dickson proved many interesting results in number theory, using results of Vinogradov to deduce the ideal Waring theorem in his investigations of additive number theory. He proved the Waring's problem for $k = 7$ $k \geq 7k \geq 7$ under the further condition of $(3k+1)/(2k-1) = [1.5k] + 1(3^k+1)/(2^k-1) \leq [1.5^k] + 1(3^k+1)/(2^k-1) \leq [1.5^k] + 1$ independently of Subbayya Sivasankaranarayana Pillai who proved it for $k = 6$ $k \geq 6k \geq 6$ ahead of him.

The three-volume History of the Theory of Numbers (1919 to 1923) is still much consulted today, covering divisibility and primality, Diophantine analysis, and quadratic and higher forms. The work contains little interpretation and makes no attempt to contextualize the results being described, yet it contains essentially every significant number theoretic idea from the dawn of mathematics up to the 1920s except for quadratic reciprocity and higher reciprocity laws. A planned fourth volume on these topics was never written. A. A. Albert remarked that this three volume work 'would be a life's work by itself for a more ordinary man.'

1.1.15 Lemma (Wallace, nilradical of modular group algebras)

Let p be a prime number, G a p-group, K a field and $char(K) = p$. $Aug(KG)$ is the nilradical of KG.

Proof. We present a proof based on an induction argument related to the order of G. Let $n \in \mathbb{N}$ and $|G| = p^n$.

Case 1: Let G be Abelian.
KG is commutative and the nilradical of von KG is exactly the set of nilpotent elements of KG. The orders of these elements of G are p-powers, and thus for all $g \in G$ we conclude – by using $char(K) = p$ and the binomial theorem – the identity $(g - 1_G)^{o(g)} = g^{o(g)} - 1_G = 0_{KG}$. Hence, $G - 1_G$ and $Aug(KG)$ are contained in $rad(KG)$. $Aug(KG)$ is a maximal ideal, and therefor the lemma is proven for case 1.

Case 2: Let G be non-Abelian.
G is a non-Abelian p-group, and thus $\{1_G\} < Z(G) < G$ is valid. By using case 1 we deduce that $Aug(KZ(G))$ is $\mathcal{A}$-nilpotent. Definition and remark 1.1.14 let us derive the $\mathcal{A}$-nilpotency of $\ker p_{Z(G)}$. Therefor, $\ker p_{Z(G)} \subseteq rad(KG)$ is valid, and $\ker p_{Z(G)}$ is contained in every maximal ideal of KG. We use induction to prove that $K(G/Z(G))$ and $KG/\ker p_{Z(G)}$ possess exactly one maximal ideal. The theorem of homomorphism let us derive that KG possess also exactly one maximal ideal. ⋄

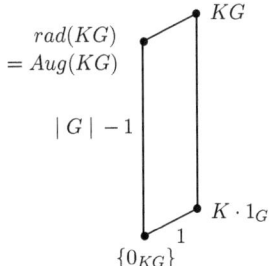

1.1.16 Remark

(i) If K is a field, then $E(K) = K \setminus \{0_K\}$ is valid, and by using part (iii) of remark 1.1.4 the identity $Q(K) = K \setminus \{-1_K\}$ is true. A straightforward calculation leads to $k' = -k(k+1_K)^{-1}$ for all $k \in K \setminus \{-1_K\}$.

(ii) Let A be an associative K-algebra and a a nilpotent element of A. An element $n \in \mathbb{N}$ exists such that $a^n = 0_A$ is true. A straightforward calculation leads to $a \in Q(A)$ and $a' = \sum_{i=1}^{n-1}(-1_K)^i a^i$. ⋄

1.1.17 Corollary (local group algebras, direct decompositions, p-group enhancement)

Let p be a prime number, G a p-group, K a field and $char(K) = p$.

(i) KG is a local K-algebra.

(ii) $(rad(KG), (K \setminus \{-1_K\}) \cdot 1_G)$ is a direct decomposition of $Q(KG)$.

(iii) $(1_G + rad(KG), (K \setminus \{0_K\}) \cdot 1_G)$ is a direct decomposition of $E(KG)$.

(iv) $Q(KG)$ resp. $E(KG)$ is the set of elements of KG such that their augmentation is not -1_K resp. not 0_K.

(v) G resp $G-1_G$ is a subgroup of $(1_G+rad(KG); \cdot)$ resp. of $(rad(KG); *)$.

(vi) If K is finite, then $\mid rad(KG) \mid = \mid 1_G+rad(KG) \mid = \mid K \mid^{|G|-1}$ is valid. In particular, $1_G + rad(KG)$ is a p-group containing G.

Proof. ad(i): The factor algebra by the nilradical of KG is – based on lemma 1.1.15 – $\mathcal{A}_1$-isomorphic to K, and thus part (i) is valid.

ad(ii): The pair $(Aug(KG), K \cdot 1_G)$ is a semidirect decomposition of KG. $K \cdot 1_G$ is central, and proposition 1.1.7 and part (i) of corollary 1.1.8 let us deduce that $(Q(Aug(KG)), Q(K \cdot 1_G))$ is a direct decomposition of $Q(KG)$. Based on lemma 1.1.15 and remark 1.1.16 part (ii) is proven.

ad(iii): This statements is a direct consequence of (ii) and part (iii) of corollary 1.1.8.

ad(iv): The function aug is a $\mathcal{A}_1$-homomorphism. Based on part (iii) of lemma 1.1.15 all elements of $E(KG)$ possess an augmentation not equal to zero. Let $x \in KG$ such that $aug(x) \neq 0_K$ is true. $(Aug(KG), K \cdot 1_G)$ is a semidirect decomposition of KG, and thus elements $k \in K$ and $r \in Aug(KG)$ exist such that $x = r + k1_G$ is true. We deduce $k \neq 0_K$. Because of $x = (1_G + k^{-1}r) \cdot (k1_G) \in (1_G + Aug(KG)) \cdot ((K \setminus \{0_K\}) \cdot 1_G)$ and part

(ii) we deduce $x \in E(KG)$. By this statement for $E(KG)$ and part (iii) of remark 1.1.4 we conclude that $Q(KG)$ is exactly the set of elements of KG possessing the augmentation -1_K.

ad(v): $G - 1_G$ is contained in $Aug(KG)$. Based on lemma 1.1.15 and part (ii) of remark 1.1.16 the set $Aug(KG)$ is a group with respect to $*$. The statement is proven by using part (iii) of remark 1.1.4.

ad(vi): This statement is a direct consequence of remark 1.1.13.◇

1.1.18 Remark

For a prime number p, a p-group G and a finite field K such that $char(K) = p$ is valid the sets G and $1_G + rad(KG)$ are identical if and only if G and K possess exactly 2 elements.◇

1.1.19 Example

Let $G := Q_8$ and K be a field with 2 elements. Based on corollary 1.1.17 the star group resp. the group of units of KG contains exactly the elements $\overline{T}$ such that T is a subset of G of even resp. uneven order. We use part (vi) of corollary 1.1.17 and conclude that these are $2^7 = 128$ elements.◇

1.1.20 Remark

If K is a field and G the trivial group, then KG is $\mathcal{A}_1$-isomorphic to K which is a local algebra.◇

1.1.21 Theorem (local group algebras)

If K is a field and G a finite, non-trivial group, then the following statements are equivalent:

(i) KG is local.

(ii) G is a p-group and $char(K) = p$ is valid.

(iii) $KG/rad(KG)$ is $\mathcal{A}_1$-isomorphic to K.

Proof. Let KG be local. By definition, KG possesses exactly one maximal ideal, and thus $Aug(KG) = rad(KG)$ is valid. If $char(K) = 0$ would be valid, then we would use the theorem of Maschke to deduce that $rad(KG) = Aug(KG)$ is the zero space. This is a contradiction to the fact $|G| \neq 1$. Hence, a prime number p exists such that $char(K) = p$ is valid. Let us assume that a prime number $q \neq p$ of $|G|$ exist. We use an element $g \in G$ such that $o(g) = q$ is valid, and we define $H := \langle g \rangle_\mathcal{G}$. Based on proposition 1.1.10 the element $e_H := \frac{1}{q_K} \sum_{i=1}^{q} g^i$ is an idempotent of KG such that $aug(e_H) = 1_K$ is valid. We deduce that $1_G - e_H$ is an idempotent of $Aug(KG) = rad(KG)$. We use the $\mathcal{A}$-nilpotency of $rad(KG)$ to deduce $1_G - e_H = 0_{KG}$ which is a contradiction. Thus, G is a p-group and part (ii) is proven. The implication $(ii) \Rightarrow (iii)$ was already proven within corollary 1.1.17. If part (iii) is valid, then statement (i) is straightforward to verify.◇

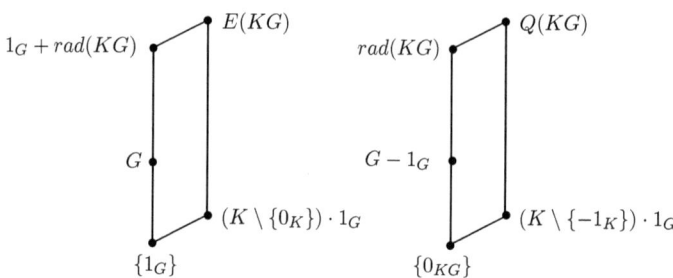

1.2 Cores

1.2.1 Definition and remark (core of a subgroup)

Let G be a group and U a subgroup of G. By $core_G(U) := \bigcap_{g \in G} U^g$ we denote the core of U in G which is the largest normal subgroup of G contained in U.◇

1.2.2 Theorem (normal subsets of G in $E(KG)$)

Let p be a prime number, G a finite p-group, K a field, $char(K) = p$ and T a subset of G. The following statements are equivalent:

(i) For all $x \in E(KG)$ the identity $T^x \subseteq G$ is valid.

(ii) For all $x \in 1_G + rad(KG)$ the identity $T^x \subseteq G$ is valid.

(iii) For all $x \in E(KG)$ the identity $T^x \subseteq T$ is valid.

(iv) For all $x \in 1_G + rad(KG)$ the identity $T^x \subseteq T$ is valid.

(v) T is central in G.

Proof. The statements (i) and (ii) as well as (iii) and (iv) are equivalent based on part (iii) of corollary 1.1.17, and it is straightforward to verify that statement (v) implies statement (iv) and that statement (iv) implies statement (ii).
Let $T^x \subseteq G$ be valid for all $x \in E(KG)$, and let us assume that U is not central in G. Elements $t \in T$ and $g \in G$ exists such that

(1) $tg \neq gt$ is true.

Case 1: Let $char(K) \neq 3$ be valid.
We define $x := 1_G + (1_G - g) + (1_G - t)$. By using theorem 1.1.21 we deduce $x \in 1_G + rad(KG)$, and based on $T^x \subseteq G$ an element $h \in G$ exist such that $tx = xh$ is true. This identity leads to

(2) $3_K t - 3_K h - tg - t^2 + gh + th = 0_{KG}$.

Because of (1) we deduce $t \notin \{t^2, tg, g, 1_G\}$. If $t = h$ would be true, then (2) would imply the identity $-hg - h^2 + 2_K gh = 0_{KG}$. But (1) implies $h^2 \neq hg \neq gh$, and this would be a contradiction. Thus, we have proven

(3) $t \notin \{t^2, tg, h, g, 1_G\}$.

Because of $char(K) \neq 3$ the identities (2) and (3) let us deduce that $t = gh$ or $t = th$ is valid.

Case 1.1: Let $t = th$ be valid which is equivalent to $h = 1_G$.
By using the equation (2) we deduce

(4) $4_K t - 3_K 1_G - tg - t^2 + g = 0_{KG}$.

Statements (3) and (4) let us conclude that $p = 2$ and $1_G + tg + t^2 + g = 0_{KG}$ are valid, and by using (1) we deduce $t^2 = 1_G$ and $tg = g$. Thus, $t = 1_G$ is valid contradicting (1).

Case 1.2: Let $t \neq th$ be valid.
We deduce $t = gh$, and by using (3) and (2) we derive $p = 2$ and

(5) $h + tg + t^2 + th = 0_{KG}$.

If $t^2 = tg$, and thus $t = g$ would be valid, then (1) is contradicted.
If $t^2 = th$, then we use (5) to deduce $t = h$ and $h = tg$. Thus, $t = tg$ and $g = 1_G$ are valid again contradicting (1). Hence, $t^2 = h$ and $tg = th$ are valid. We deduce $g = t^2$ which is again contradicting (1).

Case 2: Let $char(K) = 3$ be valid.
We define $y := 1_G + (1_G - g) + (1_G - t) + (1_G - tg)$. By using theorem 1.1.21 we deduce $y \in 1_G + rad(KG)$, and because of $T^y \subseteq G$ an element $h \in G$ exists such that $ty = yh$ is valid. We use $char(K) = 3$ to derive the equation

(6) $-tg - t^2 + t - t^2g + gh + th - h + tgh = 0_{KG}$,

and by using (1) we deduce

(7) $t \notin \{tg, t^2, t^2g\}$.

Case 2.1: Let $t = th$ be valid.
In this case $h = 1_G$ is true. We use (6) to conclude $2_K t - t^2 - t^2 g + g - 1_G = 0_{KG}$. Because of (1) the element t is not contained in the set $\{t^2, g, 1_G\}$ contradicting $p = 3$.

Case 2.2: Let $t = h$ be valid.
We use (6) to deduce $-tg - t^2g + gt + tgt = 0_{KG}$. Because of (1) the element tg is not contained in the set $\{t^2g, gt, tgt\}$ which is a contradiction.

Case 2.3: Let $h \neq t \neq th$ be valid.
By using (6), (7) and $p = 3$ we deduce $t = gh = tgh$. Thus, $t = 1_G$ is valid contradicting (1). ◇

This theorem is used to deduce one part of a theorem of K.R. Pearson:[2]

1.2.3 Corollary (normal subgroups of G in $E(KG)$)

If p is a prime number, G a finite p-group, U a subgroup of G and K a field such that $char(K) = p$ is valid, then the following statements are true:

(i) U is normal in $E(KG)$ resp. in $1_G + rad(KG)$ if and only if U is central in G.

(ii) G is normal in $E(KG)$ resp. in $1_G + rad(KG)$ if and only if G is Abelian. (Pearson [53])

Proof. These statements are a direct consequence of theorem 1.2.2. ◇

1.2.4 Corollary (cores of subgroups of G in $E(KG)$)

If p is a prime number, G a finite p-group, U a subgroup of G and K a field such that $char(K) = p$ is valid, then the following statements are valid:

(i) $core_{E(KG)}(U) = core_{1_G + rad(KG)}(U) = Z(G) \cap U \leq core_G(U)$

---

[2]K.R. Pearson dies 2015. Within the link [80] a nice tribute to him is presented.

Cores and normalizers 23

(ii) $core_{E(KG)}(G) = core_{1_G+rad(KG)}(G) = Z(G)$

(iii) $core_{E(KG)}(U) = core_{E(KG)}(G) \cap U$.

Proof. ad(i): $Z(G) \cap U$ is contained in U and is normal in $E(KG)$ and in $1_G + rad(KG)$. Based on definition and remark 1.2.1 the identity $(Z(G) \cap U) \subseteq (core_{E(KG)}(U) \cap core_{1_G+rad(KG)}(U))$ is valid. We use part (i) of corollary 1.2.3 to finalize part (i).

ad(ii)+(iii): These statements are a direct consequence of part (i).◇

1.2.5 Example

Let $G := Q_8$ and K a field such that $char(K) = 2$ is valid. Every subgroup of G is normal in G, and thus – by using definition and remark 1.2.1 – every subgroup of G is identical to its core in G. Every non-trivial subgroup U of G contains the center of G. Hence, we conclude based on part (i) of corollary 1.2.4 the identity $core_{E(KG)}(U) = core_{1_G+rad(KG)}(U) = Z(G)$.◇

1.3 Normalizers

1.3.1 Definition (set of functions)

Let M and N be sets. By $Abb(M, N)$ resp. N^M we denote the set of functions between M and N.◇

1.3.2 Remark

Let M be a set, T a subset of M and $\alpha, \beta \in Abb(M, M)$. If T is α- and β-invariant, then the identity $(\alpha\beta)|_T = \alpha|_T \beta|_T$ is valid.
The restriction to T is a homomorphism between the monoid of T-invariant functions of M in M and the monoid $Abb(T,T)$.◇

1.3.3 Proposition (enhanced group action)

Let K be a field and U a group acting on a group G by δ. For all $u \in U$ we define by $u\overline{\delta}$ the linearization of $u\delta$ on KG and $u\hat{\delta} := (u\overline{\delta})|_{E(KG)}$. U is acting by $\hat{\delta}$ on $E(KG)$, and for all $u \in U$, $g \in G$ and $k_g \in K$ the identity $(\sum_{g \in G} k_g g)(u\overline{\delta}) = \sum_{g \in G} k_g g(u\delta)$ is valid.

Proof. Let $u \in U$. $u\delta$ is a $\mathcal{G}$-automorphism of G. Hence, $u\overline{\delta}$ is a $\mathcal{A}_1$-automorphism of KG. In particular, $E(KG)$ is $u\overline{\delta}$-invariant. δ is a $\mathcal{G}$-homomorphism, and thus $\hat{\delta}$ is a $\mathcal{G}$-homomorphism, too. We deduce that U acts on $E(KG)$, and the mentioned identity is straightforward to verify.◇

1.3.4 Definition (enhanced group action)

Let K be a field and U a group acting on a G by δ. We call the action $\hat{\delta}$ defined within proposition 1.3.3 the enhanced group action on $E(KG)$ based on U with respect to δ. ⋄

Within the next lemma we analyze the connection between δ and $\hat{\delta}$ with respect to fixed points.

1.3.5 Lemma (fixed point lemma)

Let p be a prime number, K a field, $char(K) = p$ and U a p-group acting on a p-group G by δ. The following statements are equivalent:

(i) G possesses a fixed point for δ.

(ii) $E(KG)$ possesses a fixed point for $\hat{\delta}$.

Proof. The implication from (i) to (ii) is straightforward to verify. Let $x \in E(KG)$ be a fixed point of $E(KG)$ with respect to $\hat{\delta}$. Based on theorem 1.1.21 for every $g \in G$ exactly one $k_g \in K$ exists such that $x = \sum_{g \in G} k_g g$ and $0_K \neq aug(x) = \sum_{g \in G} k_g$ are valid. We use proposition 1.3.3 and deduce

(1) $\forall u \in U : \sum_{g \in G} k_g g = \sum_{g \in G} k_g g(u\delta)$.

Let $n \in \mathbb{N}$, $B_1, \ldots, B_n$ the U-orbits of G with respect to δ and $g_i \in B_i$ for all $i \in \underline{n}$. We use (1), and thus for all $i \in \underline{n}$ and $a \in B_i$ the identity $k_a = k_{g_i}$ is true, and hence we conclude

(2) $0_K \neq aug(x) = \sum_{i=1}^{n} \mid B_i \mid_K k_{g_i}$.

U is a p-group, and thus the length of each U-orbit of G with respect to δ is a p-power. By using the identity (2) and $char(K) = p$ we deduce that at least one U-orbit B_i ($i \in \underline{n}$) must exist possessing length 1. The unique element g_i of B_i is a fixed point of G with respect to δ. ⋄

We use the fixed point lemma 1.3.5 to determine the normalizer of subgroups U of G in $E(KG)$. The special case $U = G$ – which leads to the result $N_{E(KG)}(G) = G \cdot Z(E(KG))$ – was already proven by D.B. Coleman in [19]. By analyzing his proof we extend his argumentation presented in the previous fixed point lemma.

1.3.6 Theorem (normalizer of a subgroup of G in $E(KG)$)

Let p be a prime number, K a field, $char(K) = p$, G a p-group and U a subgroup of G. The identity

Cores and normalizers

$$N_{E(KG)}(U) = N_G(U) \cdot C_{E(KG)}(U)$$

is valid. In particular,

$$N_{E(KG)}(G) = G \cdot Z(E(KG))$$

is true.

Proof. Let $a \in N_{E(KG)}(U)$. For every $u \in U$ per definition the identity $u^a \in U$ is valid, and we define the function $u\delta : G \longrightarrow G, g \mapsto u^{-1}gu^a$. For every $u \in U$ the function $u\delta$ is a permutation of G. If $u, v \in U$ and $g \in G$, then the identity $g((uv)\delta) = (uv)^{-1}g(uv)^a = v^{-1}u^{-1}gu^av^a = g(u\delta)(v\delta)$ is valid. Hence, the function $\delta : U \longrightarrow S_G, u \mapsto u\delta$ is a $\mathcal{G}$-homomorphism, and for all $u \in U$ the statement

(1) $a(u\hat{\delta}) = u^{-1}au^a = a$

is true. Thus, a is a fixed point of $E(KG)$ with respect to $\hat{\delta}$, and by using lemma 1.3.5 we deduce that a fixed point g of G with respect to δ exists. Per definition of a fixed point we deduce

(2) $\forall u \in U : u^{-1}gu^a = g$.

Let $u \in U$. We use (2) and conclude $u^a = u^g \in U$. Thus, $g \in N_G(U)$ and $ag^{-1} \in C_{E(KG)}(U)$ are valid and the theorem is proven.⋄

1.3.7 Corollary

Let p be a prime number, K a field, $char(K) = p$, G a p-group and U a subgroup of G. The identities

$$N_{1_G+rad(KG)}(U) = N_G(U) \cdot C_{1_G+rad(KG)}(U)$$

and

$$N_{1_G+rad(KG)}(G) = G \cdot Z(1_G+rad(KG)) = G \cdot (Z(E(KG)) \cap (1_G+rad(KG)))$$

are valid.

Proof. Because of $N_G(U) \subseteq 1_G + rad(KG)$ the identities are straightforward to be proven using theorem 1.3.6 and Dedekind's identity.⋄

Based on corollaries 1.3.7 and 1.2.4 we are able to present some aspects of cores and normalizers:

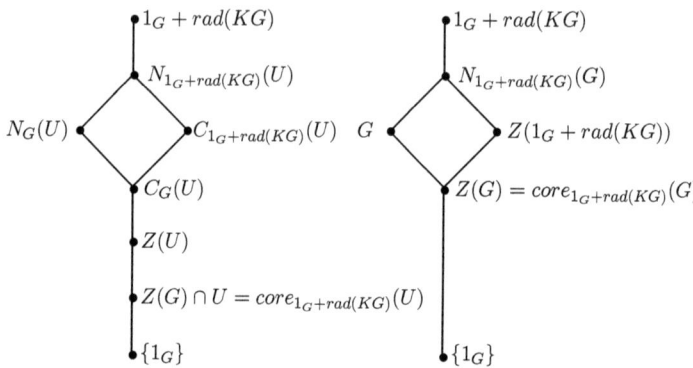

1.3.8 Corollary

Let p be a prime number, K a field, $char(K) = p$, G a p-group, U a subgroup of G and V a subgroup of $E(KG)$. The following statements are equivalent:

(i) For all $v \in V$ the identity $U^v \in G$ is valid.

(ii) V is a subset of $G \cdot C_{E(KG)}(U)$.

Proof. The implication from (ii) to (i) is straightforward to be proven. Let $U^v \in G$ for all $v \in V$ be valid. The subgroup $X := \langle U^v \mid v \in V \rangle_{\mathfrak{g}}$ is contained in G. V is a subgroup which normalizes X. Based on corollary 1.3.7 we deduce $V \leq N_{E(KG)}(X) = N_G(X) \cdot C_{E(KG)}(X)$. We use $N_G(X) \leq G$ and $C_{E(KG)}(X) \leq C_{E(KG)}(U)$ because U is contained in X. ◊

1.3.9 Corollary

Let p be a prime number, K a field, $char(K) = p$, G a p-group and T a subset of G. The following statements are valid:

(i) $N_{E(KG)}(T) = N_G(T) \cdot C_{E(KG)}(T)$

(ii) $N_{1_G+rad(KG)}(T) = N_G(T) \cdot C_{1_G+rad(KG)}(T)$

Cores and normalizers 27

(iii) The index of $N_{E(KG)}(T)$ in $N_{E(KG)}(\langle T \rangle_g)$ is identical to the index of $N_G(T)$ in $N_G(\langle T \rangle_g)$.

(iv) The index of $N_{1_G+rad(KG)}(T)$ in $N_{1_G+rad(KG)}(\langle T \rangle_g)$ is identical to the index of $N_G(T)$ in $N_G(\langle T \rangle_g)$.

Proof. ad(i): The normalizer of T in $E(KG)$ is contained in the normalizer of $\langle T \rangle_g$. We use theorem 1.3.6 to deduce that $N_{E(KG)}(T) \subseteq N_G(\langle T \rangle_g) \cdot C_{E(KG)}(\langle T \rangle_g)$ is valid. Straightforward to prove is $C_{E(KG)}(\langle T \rangle_g) = C_{E(KG)}(T)$. Thus, $N_{E(KG)}(T)$ is contained in $C_{E(KG)}(T) \cdot G$. Now let $x \in N_{E(KG)}(T)$ and let $g \in G$ and $c \in C_{E(KG)}(T)$ such that $x = c \cdot g$. We calculate $T = T^{cg} = (T^c)^g = T^g$. Therefor, $g \in N_G(T)$ is true and part (i) is proven.

ad(ii): Because of $N_G(T) \subseteq 1_G + rad(KG)$ the presented identity is straightforward to be proven using theorem 1.3.6 and Dedekind's identity.

ad(iii): We use part (i) and theorem 1.3.6 to deduce $N_{E(KG)}(T) = N_G(T) \cdot C_{E(KG)}(T)$ and $N_{E(KG)}(\langle T \rangle_g) = N_G(\langle T \rangle_g) \cdot C_{E(KG)}(\langle T \rangle_g)$. Furthermore, $N_G(T) \cap C_{E(KG)}(T) = C_G(T) = C_G(\langle T \rangle_g) = N_G(\langle T \rangle_g) \cap C_{E(KG)}(\langle T \rangle_g)$ is true. We calculate $\mid N_{E(KG)}(T) \mid = \mid N_G(T) \cdot C_{E(KG)}(T) \mid = \frac{|N_G(T)| \cdot |C_{E(KG)}(T)|}{|C_G(T)|}$ and $\mid N_{E(KG)}(\langle T \rangle_g) \mid = \frac{|N_G(\langle T \rangle_g)| \cdot |C_{E(KG)}(\langle T \rangle_g)|}{|C_G(T)|}$. From this calculation we conclude (i).

ad(iv): The argumentation is similar as done for (iii) by using part (ii) and corollary 1.3.7. ⋄

1.3.10 Definition (commutator, class number)

Let G be a group and $g, h \in G$.

(i) By $[g, h] := g^{-1}h^{-1}gh$ we denote the commutator of g with h.

(ii) If G is finite, then $c(G)$ the number of conjugacy classes in G – also called class number of G. ⋄

Another consequence of theorem 1.1.21 is the following result:

1.3.11 Proposition (dimension of the center)

Let p be a prime number, G a p-group, K a field and $char(K) = p$.

(i) $Z(rad(KG)) = Z(rad(KG)^*) = Z(KG) \cap rad(KG)$

(ii) $Z(rad(KG)) = rad(KZ(G)) \oplus_K \langle \{\overline{g^G} \mid g \in G \setminus Z(G)\} \rangle_K$
In particular, $dim_K(Z(rad(KG))) = c(G) - 1$ is valid.

(iii) If K is finite, then $\mid Z(rad(KG)) \mid = \mid K \mid^{c(G)-1}$ is true.⋄

1.3.12 Example

Let $G := Q_8$ and K a field with two elements. Based on theorem 1.1.21 the identities $E(KG) = 1_G + rad(KG)$ and $\mid 1_G + rad(KG) \mid = 2^7$ are valid. G possesses exactly three non-central conjugacy classes, and $\mid Z(G) \mid = 2$ is true. Proposition 1.3.11 leads to $\mid Z(1_G + rad(KG)) \mid = 2^4$.

The normalizer of G in $1_G + rad(KG)$ is – based on corollary 1.3.7 – a normal subgroup of $1_G + rad(KG)$ of order 2^6.

If U is a maximal subgroup of G, then U is normal and self-centralizing in G, and based on corollary 1.3.7 we deduce $\mid N_{1_G+rad(KG)}(U) \mid = 2 \cdot \mid C_{1_G+rad(KG)}(U) \mid \geq 2^5$.

If $\mid N_{1_G+rad(KG)}(U) \mid = 2^5$ would be valid, then $Z(1_G + rad(KG))$ and $C_{1_G+rad(KG)}(U)$ would be of the same order, and hence they would be identical. U is Abelian, and thus U would be central in G which is a contradiction.

If $\mid N_{1_G+rad(KG)}(U) \mid = 2^7$ would be true, then U would be a normal subgroup of $1_G + rad(KG)$. U is non-central in G contradicting corollary 1.2.4. We deduce $\mid N_{1_G+rad(KG)}(U) \mid = 2^6$, and thus $N_{1_G+rad(KG)}(U)$ is a normal subgroup of index 2 in $1_G+rad(KG)$. Therefor the identities $\mid C_{1_G+rad(KG)}(U) \mid = 2^5$, $C_{1_G+rad(KG)}(U) = U \cdot Z(1_G + rad(KG))$ and $N_{1_G+rad(KG)}(U) = G \cdot Z(1_G + rad(KG))$ are valid.

If $g \in G \backslash Z(G)$, then g possesses in $1_G+rad(KG)$ exactly 4 and in G exactly 2 conjugate elements (by focussing on the subgroup generated by g).⋄

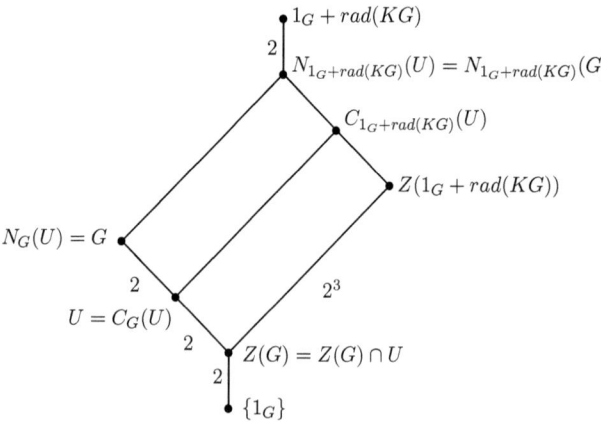

Cores and normalizers

We close this chapter by presenting a description of the centralizer of subgroups U of G in $E(KG)$. Therefor we are able to count the conjugates of elements and of subgroups of G in $E(KG)$.

1.3.13 Definition (centralizer of enhanced group action)

Let U be a group acting on a set M by δ and K be a field. For every $u \in U$ let $\overline{u\delta}$ the linearization of $u\delta$ on KM, and we define $C_{KM,\delta}(U) := \bigcap_{u \in U} ker(\overline{u\delta} - id_{KM})$. This set is called the centralizer of U in KM with respect to δ. ⋄

1.3.14 Proposition (basis of the centralizer)

Let U be a group acting on a finite set M by δ, $B_1, \ldots, B_n$ the U-orbits of M and K a field. The set $\{\overline{B_i} \mid i \in \underline{n}\}$ is a K-basis of the K-subspace $C_{KM,\delta}(U)$ of KM.

Proof. Based on definition 1.3.13 we deduce that $C_{KM,\delta}(U)$ is a K-subspace of KM. If $x \in KM$, then for every $i \in \underline{n}$ and for every $b \in B_i$ exactly one $k_b \in K$ exists such that $x = \sum_{i=1}^{n} \sum_{b \in B_i} k_b b$ is valid. For all $u \in U$ we deduce that $x(\overline{u\delta}) = x$ is true if and only if

(1) $\sum_{i=1}^{n} \sum_{b \in B_i} k_b b(u\delta) = \sum_{i=1}^{n} \sum_{b \in B_i} k_b b$

is valid. U is acting transitive on every U-orbit of M, and thus the proof is finished by using (1). ⋄

1.3.15 Definition (conjugation)

If G is a group, then we use the symbol $Inn(G)$ for the group of inner automorphism of G and we define for all $g \in G$ the function $\kappa_g : G \longrightarrow G$, $x \mapsto x^g$ and, in addition, $\kappa : G \longrightarrow Inn(G)$, $g \mapsto \kappa_g$. ⋄

1.3.16 Corollary (properties of the centralizer)

Let p be a prime number, K a field, $char(K) = p$, G a finite p-group, U a subgroup of G and $\mathcal{B}$ the set of U-orbits of G with respect to $\kappa_{|_U}$. The following statements are valid:

(i) The set $\{\overline{B} \mid B \in \mathcal{B}\}$ is a K-basis of the K-space $C_{KG}(U)$.

(ii) $C_{rad(KG)}(U) = Aug(KC_G(U)) \oplus_K \langle \{\overline{B} \mid B \in \mathcal{B}, \mid B \mid \neq 1\} \rangle_K$

(iii) $C_{1_G + rad(KG)}(U) = 1_G + C_{rad(KG)}(U)$

(iv) If K is finite, then $\mid C_{1_G+rad(KG)}(U) \mid = \mid K \mid^{\mid \mathcal{B} \mid -1}$ is valid.

Proof. ad(i): This statement is a direct consequence of proposition 1.3.14.

ad(ii): Based on (i) the sum presented is direct. We use theorem 1.1.21 to conclude $rad(KG) = Aug(KG)$, and thus the sum presented is contained in $C_{rad(KG)}(U)$. Because of $1_G \notin rad(KG)$ a dimension argument proves (ii).

ad(iii): This statement is straightforward to prove.

ad(iv): This statement is a direct consequence of (i) and (iii). ⋄

1.3.17 Corollary (conjugates and conjugate subgroups)

Let p be a prime number, K a finite field, $char(K) = p$ and G a p-roup. The following statements are valid:

(i) If $g \in G \backslash Z(G)$ and $\mathcal{B}$ is the set of $\langle g \rangle_g$-orbits of G with respect to $\kappa_{\mid \langle g \rangle_g}$, then g possesses exactly $\mid K \mid^{\mid G \mid - \mid \mathcal{B} \mid}$ conjugates in $1_G + rad(KG)$.

(ii) If U is a subgroup of G and $\mathcal{B}$ the set of U-orbits of G with respect to $\kappa_{\mid U}$, then $\mid N_{1_G+rad(KG)}(U) \mid = \frac{\mid N_G(U) \mid}{\mid C_G(U) \mid} \cdot \mid K \mid^{\mid \mathcal{B} \mid -1}$ is valid.
In particular, $1_G + rad(KG)$ possesses exactly $\frac{\mid C_G(U) \mid}{\mid N_G(U) \mid} \cdot \mid K \mid^{\mid G \mid - \mid \mathcal{B} \mid}$ subgroups conjugated to U.

Proof. ad(i): This statement is a direct consequence of corollary 1.3.16.

ad(ii): This statement is a direct consequence of part (iv) of corollary 1.3.16 and corollary 1.3.7. ⋄

The fixed point lemma can be used to prove that g^G is exactly the intersection of $g^{E(KG)}$ with G. In other words, the conjugacy classes of G can be enhanced to conjugacy classes in $E(KG)$ uniquely. This is a theorem of Coleman which is discussed within the exercises 18 and 19. If $g \in G \backslash Z(G)$, then the length of the class of g in $1 + rad(KG)$ is exactly $\mid K \mid^{\mid \mathcal{B} \mid -1}$ (see part (i) of corollary 1.3.17).

1.3.18 Example

Let $G := Q_8 = \{1_G, i^2, i, j, k, i^{-1}, j^{-1}, k^{-1}\}$ and K a field with 2 elements. The $\langle i \rangle_g$-orbits of G under $\kappa_{\mid \langle i \rangle_g}$ are $\{1_G\}$, $\{i^2\}$, $\{i\}$, $\{i^{-1}\}$, $\{j, j^{-1}\}$ and $\{k, k^{-1}\}$. Based on corollary 1.3.17 we deduce that $1_G + rad(KG)$ possesses exactly 4 conjugates of i and exactly 2 conjugate subgroups to $\langle i \rangle_g$ (see example 1.3.11). A straightforward calculation leads to
$i^1 = i$, $i^j = i^{-1}$, $i^{1_G+i+j} = i^3 + \overline{j^G} + \overline{k^G}$ and $i^{1_G+j+k} = (i^3)^{1_G+i+j} =$

Cores and normalizers

$i + \overline{j}^G + \overline{k}^G$. Thus,
$i^{1_G + rad(KG)} = \{i, i^3, i^{1_G+i+j}, i^{1_G+j+k}\}$ and
$\langle i \rangle_{\mathfrak{g}}^{1_G+i+j} = \{1, i^2, i^{1_G+i+j}, (i^3)^{1_G+i+j}\} \neq \langle i \rangle_{\mathfrak{g}}$ are true.⋄

1.4 Open topics and exercises

Let K be a field, $char(K) = p$, G a finite p-group and S a subset of G.

Open-ended questions 1 *(i) Determine the conjugacy classes of $1 + rad(KG)$ which have trivial intersection with G. Determine their quantity and length.*

(ii) Analyze the inner structure of the normalizer and centralizer of a subgroup U of G in $E(KG)$. The structure of the center – which arises from the special case $U = G$ – will be analyzed within chapters 3 and 4.

Excercise 1 *Let $g \in G \setminus Z(G)$. Compare the length of the class of g in G with the one of g in $1 + rad(KG)$.*

Excercise 2 *Let U be a subgroup of G. Prove that $core_G(U) = core_{E(KG)}(U)$ is valid if and only if U is central in G.*

Excercise 3 *Determine the connection between the centralizer of S and of $\langle S \rangle_{\mathfrak{g}}$ in G.*

Excercise 4 *Do a research in the literature and define the defect of subnormality of a subgroup within the group and find equivalent descriptions.*

Excercise 5 *Prove for a nilpotent group that the defect of subnormality for each subgroup can be bounded by the class of nilpotency of G. Deduce that the defect of subnormality of G in $E(KG)$ is not greater than $cl(1+rad(KG)) = cl(rad(KG)°)$. Use the theorem of Du for radical rings.*

Excercise 6 *Prove remark 1.1.2 in details.*

Excercise 7 *Let L be a Lie algebra and $a \in L$. Define $at(a)$ by $bat(a) := ab$ for all $b \in L$. Analyze the characteristics of $at(a)$ and the connection to $ad(a)$ defined by $bad(a) := ab$ for all $b \in L$. What are the connections between the maps $ad(.)$ and $at(.)$?*

Excercise 8 *Prove remark 1.1.4 in details.*

Excercise 9 *Let K be a field and $n \in \mathbb{N}$. Determine the units and quasi regular elements of K, $\mathbb{C}$, $\mathbb{H}$, $K^{n \times n}$, the set of lower triangular matrices of $K^{n \times n}$ and of the set of strict lower triangular matrices of $K^{n \times n}$.*

Excercise 10 *Is every nilpotent element of an associative algebra quasi regular? Is the opposite statement true, too?*

Excercise 11 *Let K be a field and $n \in \mathbb{N}$. Present the group of units resp. the quasi regular group of the set of lower triangular matrices over a field by using the semidirect decomposition of the algebra as sum of diagonal and strict lower triangular matrices.*

Excercise 12 *Let $p \neq 2, 3$, $K := GF(5)$ and $G := S_3$. For every subset T of S_3 calculate e_T and determine whether e_T is an idempotent. On what terms are these elements resp. idempotent elements central? Do the same exercise by using $GF(p)$ instead of $GF(2)$ and $GF(3)$.*

Excercise 13 *Let $n \in \mathbb{N}$. Determine n_K in $GF(p)$ and $GF(p^k)$ for all $k \in \mathbb{N}_{\geq 2}$.*

Excercise 14 *Let K be a field, G a finite group such that KG is semisimple and H a subset of G. On what terms on e_H is H a normal subgroup?*

Excercise 15 *Let K be a field, G a finite group, KG semisimple and H a subset of G. On what terms on e_H is H a normal subset?*

Excercise 16 *Let K be a field, G a finite group, KG and H a subset of G. On what terms on e_H is H a subgroup?*

Excercise 17 *Apply theorem 1.3.6 to a normal subgroup U of G.*

Excercise 18 *Apply the fixed point lemma 1.3.5 and the argumentation of theorem 1.3.6 to prove the following theorem of D.B. Coleman presented in [19]: Let G be a finite p-group, K a field, $char(K) = p$, $g \in G$ and $x \in E(KG)$. If $[g, x] \in G$ is valid, then an element $h \in G$ exists such that $[g, x] = [g, h]$ is true. (Hint: use the subgroup in G generated by g.)*

Excercise 19 *Under the assumptions of exercise 18 prove: If $x^{E(KG)} \cap G \neq \emptyset$, then an element $h \in G$ exists such that $x^{E(KG)} \cap G = h^G$ is valid. In particular, $g^{E(KG)} \cap G = g^G$ is true. Every conjugacy class of G can be enhanced to exactly one conjugacy class of $E(KG)$. Thus, $c(E(KG)) \geq c(G)$ is valid.*

Excercise 20 *Let G be a finite p-group and K a field of characteristic p. By using theorem 1.3.6 analyze the connection between $\mid K \mid^{|G|-c(G)}$ and $\frac{|G|}{|Z(G)|}$.*

Excercise 21 *Let $K := GF(2)$, $G := S_3$ and $A := KG$. For every subset T of G determine $\langle T \rangle_K$, $\langle T \rangle_A$ and $\langle T \rangle_{A_1}$ as well as a basis and the dimension* of these K-subspaces. Calculate the ideal generated by T in A. Calculate the ideal and the subalgebra generated by T in the associated Lie algebra A° of A with respect to the multiplication $a \circ b := ab - ba$. Present the results by using a Hasse diagram!

Excercise 22 Let K be a field and $G := S_4$. For every normal subgroup N of G focus on the map p_N and calculate a basis and the dimension of its kernel and its image. For which fields K is the kernel resp. the image of p_N nilpotent, semisimple or separable? Present the results by using a Hasse diagram!

Excercise 23 Let $K := GF(2)$ and $G := S_3$. Determine the representative matrix of the linear function aug based on the basis G and $\{1\}$. Calculate all elements which are mapped by aug to 1 resp. to 0. Present the results by using a Hasse diagram!

Excercise 24 Let K be a field and G a finite cyclic group generated by g. For the left and right multiplication on KG with g determine the representative matrix based on the basis G. Is this matrix invertible? If possible determine the inverse of this matrix. Calculate the determinant of the matrix and of its inverse. In what way is the inverse of the matrix related to g^{-1}? Present the results by using a Hasse diagram! Is it possible to use the method within this exercise to calculate the inverse in KG for an arbitrary element? Is the method restricted to a cyclic group?

Excercise 25 Let K be a field, $char(K) = 2$ and G a quaternion, dihedral or semi-dihedral 2-group. Which subgroups of G are normal in $E(KG)$? Determine the core of an arbitrary subgroup of G in $E(KG)$! Present the results by using a Hasse diagram!

Excercise 26 Let K be a field, $char(K) = 2$ and G a quaternion, dihedral or semi-dihedral 2-group. Which subsets of G are normal in $E(KG)$? Which subsets T of G possess the characteristic $T^x \in G$ for all $x \in E(KG)$? Are those subsets the same as the normal ones and why? Present the results by using a Hasse diagram!

Excercise 27 Let $K := GF(2)$ and $G := D_8$. Determine the nilpotency class of $rad(KG)$ and the maximal nilpotency class of the elements of $rad(KG)$. Are these two numbers identical? Do elements $x \in rad(KG)$ exist such that $cl(x) < cl(rad(KG))$ is valid?

Excercise 28 Let $K := GF(2)$ and $G := D_8$. Determine the order of $1 + rad(KG)$. In what way is this order a bound for the orders of the elements of $1 + rad(KG)$? Does an element $x \in rad(KG)$ exist such that $o(x) = \mid 1 + rad(KG) \mid$ is valid? Does an element $x \in rad(KG)$ exist such that $o(x) < \mid 1 + rad(KG) \mid$ is true? Present the results by using a Hasse diagram!

Excercise 29 Let $K := GF(2)$ and $G := D_8$. Prove that at least 5 conjugacy classes in $1 + rad(KG)$ exist. How many classes do exist? What is their size? Present the results by using a Hasse diagram!

Excercise 30 Let A be an associative K-algebra, $char(K) = p \in \mathbb{P}$, $k \in \mathbb{N}$ and $a, b \in A$ such that $ab = ba$ is valid. Prove the identity $(a+b)^{p^k} = a^{p^k} + b^{p^k}$. Discuss the importance of this identity within the lemma of Wallace (Tip: binomial theorem; almost all coefficients are divided by p; begin with the case $k = 1$ and finalize the proof by an induction argument)

Excercise 31 Let K be a finite field and G a finite group. For the following combinations of K and G determine a basis and the dimension of $rad(KG)$:

(i) $|G| = 3^4$, $char(K) = 3$

(ii) $|G| = 3^4$, $char(K) = 7$

(iii) $|G| = 2^5$, $char(K) = 2$

(iv) $|G| = 2^5$, $char(K) \neq 2$

(v) $|G| = 5^4$, $char(K) = 5$

(vi) $|G| = 5^4$, $char(K) = 3$

(vii) $|G| = 8^3$, $char(K) = 2$

(viii) $|G| = 8^3$, $char(K) = 11$.

Count the elements of $rad(KG)$ and decide whether KG is local. Decide whether $1 + rad(KG)$ and $E(KG)$ are identical. Present the results by using a Hasse diagram!

Excercise 32 Let K be a field, $char(K) = p$ and G a finite p-group. Does a subgroup U of G exist such that $N_{E(KG)}(U)$ is contained in G?

Excercise 33 Let K be a field, $char(K) = p$ and G a finite p-group of order p^3. Determine the cores of all subgroups of G in $E(KG)$. Present the results by using a Hasse diagram!

Excercise 34 Let K be a field, $char(K) = p$ and G a finite p-group. Determine the cores of the subgroups U of G in $E(KG)$. What is the connection to the cores of U in G? Is $core_{E(KG)}(U) \cap G = core_G(U)$ valid? Present the results by using a Hasse diagram!

Excercise 35 Prove remark 1.3.2 in details.

Cores and normalizers

Excercise 36 Let $K := GF(2)$ and $G := D_8$. For every subgroup U of G let us focus on the action of U on G by conjugation, right and left multiplication and of the enhanced actions of U on $E(KG)$. What is the meaning and the connection of the fixed points of these six actions? Present the results by using a Hasse diagram!

Excercise 37 Let K be a field, $char(K) = p$ and G a finite p-group. For a normal subgroup N of G let us focus on the action from G on N resp. from N on G by conjugation and on the enhanced group action from N on $E(KG)$ and from G on $E(KN)$. What is the meaning and the connection of the fixed points of these four actions? Present the results by using a Hasse diagram!

Excercise 38 Let K be a field, $char(K) = 2$ and $G \in \{Q_8, D_8, SD_8\}$. For every subgroup U of G determine $N_{E(KG)}(U)$ and $core_{E(KG)}(U)$. On what terms is U normal in $E(KG)$? Present the results by using a Hasse diagram!

Excercise 39 Prove proposition 1.3.11 is details. In addition, let $n \in \mathbb{N}$, K be finite and p be a prime number. Apply proposition 1.3.11 to the following groups and fields:

(i) $|G| = 3^3$, $char(K) = 3$

(ii) $|G| = p^3$, $char(K) = p$

(iii) $G = Q_8$, $char(K) = 2$

(iv) $G = D_8$, $char(K) = 2$

(v) $G = SD_8$, $char(K) = 2$

(vi) $G = Q_{2^n}$, $char(K) = 2$

(vii) $G = D_{2^n}$, $char(K) = 2$

(viii) $G = SD_{2^n}$, $char(K) = 2$

(ix) $|G'| = p$, $char(K) = p$.

Present the results within a Hasse diagram!

Excercise 40 In what way is part (ii) of corollary 1.3.17 useable for deducting a bound for part (i) of the same corollary?

Excercise 41 Execute example 1.3.18 for D_8 and SD_8.

Excercise 42 Execute example 1.3.18 for a group of order p^3.

Excercise 43 Let K be a field, $char(K) = p$ and G a finite p-group. The following statements are valid:

(i) Let $\alpha \in Aut(G)$. The function $\hat{\alpha} : E(KG) \to E(KG)$, $\sum_{g \in G} k_g g \mapsto \sum_{g \in G} k_g (g\alpha)$ is an automorphism of $E(KG)$.

(ii) If $\alpha \in Inn(G)$, then $\hat{\alpha} \in Inn(E(KG))$.

(iii) If $\alpha \in Aut(G) \setminus Inn(G)$, then $\hat{\alpha} \in Aut(E(KG)) \setminus Inn(E(KG))$. (Tip: Use the determination of the normalizer of G in $E(KG)$. This is a theorem of Bovdi. Do a research in the literature to find this theorem.)

(iv) The function $\hat{} : Aut(G) \longrightarrow Aut(E(KG))$ is a monomorphism.

(v) $\hat{}(Aut(G)) \cap Inn(E(KG)) = Inn(G)$

(vi) Every outer automorphism of G is mapped to an outer automorphism of $Aut(E(KG))$ by $\hat{}$.

(vii) Every inner automorphism of G is mapped to an inner automorphism of $Aut(E(KG))$ by $\hat{}$.

(viii) $Aut(G)/Inn(G)$ is isomorphic to a subgroup of $Aut(E(KG))/Inn(E(KG))$.

Chapter 2

End-commutable orderings and exponents

2.1 Basic properties

2.1.1 Definition (end-commutable ordering)

Let A be a K-algebra, $n \in \mathbb{N}$ and $a_i \in A$ for all $i \in \underline{n}$. The n-tuple $(a_1, \ldots, a_n)$ is called end-commutable if

$$\forall i \in \underline{n-1} : a_i \left(\sum_{j=i+1}^{n} a_j \right) = \left(\sum_{j=i+1}^{n} a_j \right) a_i$$

is valid. If T is a finite and non-empty subset of A, then we call a $|T|$-tuple $(a_1, \ldots, a_{|T|})$ over T an end-commutable ordering of T (with respect to the composition of A), if the tuple is end-commutable and $T = \{a_1, \ldots, a_{|T|}\}$ is valid. In this case we say that T is end-commutable. By $EA(T)$ we symbolize all end-commutable orderings of T.⋄

2.1.2 Examples

(i) Let A be a K-algebra, $n \in \mathbb{N}$, $\alpha \in S_n$ and $a_i \in A$ for all $i \in \underline{n}$. If $a_1, \ldots, a_n$ are pairwise commutating, then $(a_{1\alpha}, \ldots, a_{n\alpha})$ is end-commutable.

(ii) Let A a be a K-algebra, $char(K) = 2$ and $a, b \in A$ such that $ab \neq ba$ is valid. The 3-tuple (a, b, b) is end-commutable but $\{a, b, b\} = \{a, b\}$ possesses no end-commutable ordering.

(iii) Let K be a field, $G := D_{16}$, $a, b \in G$, $G = \langle a, b \rangle_{\mathfrak{G}}$, $o(a) = 8$, $o(b) = 2$ and $ba = a^{-1}b$. $\{ab, a^3b, a^5b, a^7b\}$ is a conjugacy class of G. The tuple (ab, a^5b, a^3b, a^7b) is an end-commutable ordering for $(ab)^G$ within KG (see part (i) of the construction 2.2.4) but not the tuple (ab, a^3b, a^5b, a^7b).⋄

2.1.3 Remark (end-commutable ordering within group algebras)

Let K be a field, G a group, T a finite and non-empty subset of G and $(t_1, \ldots, t_{|T|})$ an end-commutable ordering of T within KG. G is a K-basis of the K-space KG, and thus the tuple is end-commutable if and only if for all $i \in \underline{|T|-1}$ the sets $\{(t_{i+1})^{t_i}, \ldots, (t_{|T|})^{t_i}\}$ and $\{t_{i+1}, \ldots, t_{|T|}\}$ are identical: t_i is normalizing the set $\{t_{i+1}, \ldots, t_{|T|}\}$. Thus, we can decide within G whether an end-commutable ordering for T exists. Therefor we use the terminology end-commutable for group algebra without mentioning the field K resp. the group algebra KG. ⋄

2.1.4 Remark

Let A be a K-algebra, $n \in \mathbb{N}$ and $a_i \in A$ for all $i \in \underline{n}$. $(a_1, \ldots, a_n)$ is end-commutable if and only if for all $i \in \underline{n}$ the tuple $(a_i, \ldots, a_n)$ is end-commutable. In particular, the elements a_{n-1} and a_n are commuting. ⋄

The next theorem for end-commutable orderings is one important instrument for determining the exponent of the center of $1_G + rad(KG)$.

2.1.5 Theorem (p-powers of end-commutable orderings)

Let A be an associative K-algebra, p a prime number, $char(K) = p$, $n \in \mathbb{N}$, $a_i \in A$ for all $i \in \underline{n}$ and $(a_1, \ldots, a_n)$ end-commutable. The following statements are valid:

(i) For all $s \in \mathbb{N}$ the identity $(\sum_{i=1}^{n} a_i)^{p^s} = \sum_{i=1}^{n} a_i^{p^s}$ is true.

(ii) For all $s \in \mathbb{N}$ the tuple $(a_1^{p^s}, \ldots, a_n^{p^s})$ is end-commutable.

Proof. We begin the proof by remarking that for all $x, y \in A$ such that $xy = yx$ is valid – because of $char(K) = p$ and the binomial theorem – the identity

(1) $(x+y)^p = x^p + y^p$

is true. The statements (i) and (ii) are proven at first for the case $s = 1$.

ad(i): The case $n = 1$ is straightforward to prove. By definition a_1 and $\sum_{i=2}^{n} a_i$ are commuting, and thus we use (1) to deduce $(\sum_{i=1}^{n} a_i)^p = a_1^p + (\sum_{i=2}^{n} a_i)^p$. By using remark 2.1.4 an induction argument proves statement (i).

ad(ii): Let $i \in \underline{n-1}$. We deduce

$$a_i{}^p \left(\sum_{j=i+1}^{n} a_j{}^p \right) \qquad \text{(see (ii) and remark 2.1.4)}$$

$$= a_i{}^p \left(\sum_{j=i+1}^{n} a_j \right)^p \qquad \text{(see (1) and definition 2.1.1)}$$

$$= \left(a_i \left(\sum_{j=i+1}^{n} a_j \right) \right)^p \qquad \text{(see definition 2.1.1)}$$

$$= \left(\left(\sum_{j=i+1}^{n} a_j \right) a_i \right)^p \qquad \text{(see (1) and definition 2.1.1)}$$

$$= \left(\sum_{j=i+1}^{n} a_j \right)^p a_i{}^p \qquad \text{(see (ii) and remark 2.1.4)}$$

$$= \left(\sum_{j=i+1}^{n} a_j{}^p \right) a_i{}^p.$$

Hence, the statement (ii) is true for $s = 1$, and a straightforward induction argument completes the proof.◇

2.1.6 Example

Let K be a field, $char(K) = 2$, $G := D_{16}$, $a, b \in G$, $G = \langle a, b \rangle_{\mathfrak{g}}$, $o(a) = 8$, $o(b) = 2$ and $ba = a^{-1}b$. $C := \{ab, a^3b, a^5b, a^7b\}$ is a conjugacy class of G. Based on example 2.1.2 the tuple (ab, a^5b, a^3b, a^7b) is an end-commutable ordering of C. We apply theorem 2.1.5 and calculate $(ab + a^3b + a^5b + a^7b)^2 = (ab)^2 + (a^3b)^2 + (a^5b)^2 + (a^7b)^2$. All elements of C are involutions, and thus $(\overline{C})^2 = 0_{KG}$ is true: $\overline{C}$ is an involution.◇

We want to prove that every finite group is nilpotent if and only if every normal subset possesses an end-commutable ordering. For this proof we need the following properties of end-commutable orderings.

2.1.7 Proposition

Let A be a K-algebra, $n \in \mathbb{N}$ and $a_i \in A$ for all $i \in \underline{n}$. The following statements are valid:

(i) If α is a K-algebra endomorphism of A and $(a_1, \ldots, a_n)$ is end-commutable, then $(a_1\alpha, \ldots, a_n\alpha)$ is end-commutable.

(ii) Let $i \in \underline{n}$ and $a_i \in C_A(\{a_1, \ldots, a_n\})$. $(a_1, \ldots, a_n)$ is end-commutable if and only if $(a_1, \ldots, a_{i-1}, a_{i+1}, \ldots, a_n)$ is end-commutable.

(iii) Let $r \in \mathbb{N}$, $b_i \in A$ for all $i \in \underline{r}$ and $(a_1, \ldots, a_n)$, $(b_1, \ldots, b_r)$ end-commutable. If for all $i \in \underline{n}$ the identity $a_i \left(\sum_{j=1}^{r} b_j \right) = \left(\sum_{j=1}^{r} b_j \right) a_i$ is valid, then $(a_1, \ldots, a_n, b_1, \ldots, b_r)$ is end-commutable.

(iv) Let A be associative, $r \in \mathbb{N}$, $b_i \in A$ for all $i \in \underline{r}$ and $(a_1, \ldots, a_n)$, $(b_1, \ldots, b_r)$ be end-commutable. If for all $i \in \underline{n}$ and for all $j \in \underline{r}$ the identities $a_i b_j = b_j a_i$ and $b_j (\sum_{t=1}^{r} b_t) = (\sum_{t=1}^{r} b_t) b_j$ are valid, then $(a_1 b_1, \ldots, a_1 b_r, \ldots, a_n b_1, \ldots, a_n b_r)$ is end-commutable.

Proof. ad(i): Let $i \in \underline{n-1}$. The following argumentation is valid:

$$(\sum_{j=i+1}^{n} a_j \alpha) a_i \alpha = ((\sum_{j=i+1}^{n} a_j) a_i) \alpha = (a_i (\sum_{j=i+1}^{n} a_j)) \alpha = a_i \alpha (\sum_{j=i+1}^{n} a_j \alpha).$$

ad(ii): This statement is straightforward to prove.

ad(iii): Let $i \in \underline{n}$. The following calculation is valid:

$$a_i ((\sum_{j=i+1}^{n} a_j) + (\sum_{s=1}^{r} b_s)) = a_i (\sum_{j=i+1}^{n} a_j) + a_i (\sum_{s=1}^{r} b_s) =$$
$$= (\sum_{j=i+1}^{n} a_j) a_i + (\sum_{s=1}^{r} b_s) a_i = ((\sum_{j=i+1}^{n} a_j) + (\sum_{s=1}^{r} b_s)) a_i.$$

$(b_1, \ldots, b_r)$ is end-commutable, and thus (iii) is true.

ad(iv): Let $x := \sum_{i=1}^{r} b_i$, $i \in \underline{n}$ and $j \in \underline{r}$. The following argumentation is valid:

$$(a_i b_j) ((\sum_{s=j+1}^{r} a_i b_s) + (\sum_{t=i+1}^{n} a_t x)) = b_j (\sum_{s=j+1}^{r} b_s) a_i^2 + a_i (\sum_{t=i+1}^{n} a_t) b_j x =$$
$$= (\sum_{s=j+1}^{r} b_s) b_j a_i a_i + (\sum_{t=i+1}^{n} a_t) a_i b_j x = ((\sum_{s=j+1}^{r} a_i b_s) + (\sum_{t=i+1}^{n} a_t x)) (a_i b_j). \diamond$$

2.1.8 Corollary

Let G be a finite group.

(i) If U is a subgroup of $Aut(G)$, $\alpha = (a_1, \ldots, a_n) \in EA(G)$ and $\gamma \in U$, then $\alpha\gamma := (a_1\gamma, \ldots, a_n\gamma) \in EA(G)$ is true. In particular, U is acting on $EA(G)$, and for all $\alpha \in EA(G)$ the identity $Stab_U(\alpha) = \{id_G\}$ is valid.

(ii) If U is a subgroup of $Aut(G)$ and n the number of the $EA(G)$-orbits under U, then $\mid EA(G) \mid = n \mid U \mid$ is true. In particular, $\mid Aut(G) \mid$ is a divisor of $\mid EA(G) \mid$.

Proof. ad(i): Let U be a subgroup of $Aut(G)$. Based on part (i) of proposition 2.1.7 the group U acts on $EA(G)$ as presented. If $(g_1, \ldots, g_{|G|})$ is an end-commutable ordering of G, then for all $\gamma \in U$ the condition $(g_1, \ldots, g_{|G|})\gamma = (g_1, \ldots, g_{|G|})$ is valid if and only if for all $g \in G$ the identity

$g\gamma = g$ is true.

ad(ii): Based on the fixed point theorem of Burnside[1] the identity $n = \frac{1}{|U|} \sum_{t \in EA(G)} | \, Stab_U(t) \, |$ is valid, and by using (i) the proof is completed. ◇

2.1.9 Example

Let K be a field and $G := Q_8 = \{1_G, i^2, i, j, k, i^{-1}, j^{-1}, k^{-1}\}$. Based on proposition 2.1.7 the determination of $EA(G)$ can be derived from the determination of the end-commutable orderings of $T := \{i, j, k, i^{-1}, j^{-1}, k^{-1}\}$. Let $(x_1, \ldots, x_6) \in EA(T)$. Based on definition of the end-commutable ordering the elements x_5 and x_6 are commutable, and thus $\{x_5, x_6\}$ is a conjugacy class of G. In particular, $x_5 + x_6$ is a central element of KG. We deduce $(x_4 + x_5 + x_6)^{x_3} = x_4^{x_3} + x_5 + x_6$, and thus x_3 and x_4 are commutable. Therefor $\{x_3, x_4\}$ and – based on a similar argumentation – $\{x_1, x_2\}$ are conjugacy classes of G.

If x_1, x_2 and x_3, x_4 and x_5, x_6 are distinct and pairwise commuting elements

---

[1] William Burnside (2 July 1852 to 21 August 1927) was an English mathematician. He is known mostly as an early researcher in the theory of finite groups. Burnside was born in London, and attended St. John's and Pembroke Colleges at the University of Cambridge, where he was the Second Wrangler in 1875. He lectured at Cambridge for the following ten years, before being appointed professor of mathematics at the Royal Naval College in Greenwich. While this was a little outside the main centers of British mathematical research, Burnside remained a very active researcher, publishing more than 150 papers in his career. Burnside's early research was in applied mathematics. This work was of sufficient distinction to merit his election as a fellow of the Royal Society in 1893, though it is little remembered today. Around the same time as his election his interests turned to the study of finite groups. This was not a widely studied subject in Britain in the late 19th century, and it took some years for his research in this area to gain widespread recognition. The central part of Burnside's group theory work was in the area of group representations, where he helped to develop some of the foundational theory, complementing, and sometimes competing with, the work of Ferdinand Georg Frobenius, who began his research in the subject during the 1890s. One of Burnside's best known contributions to group theory is his $p^a q^b$ theorem, which shows that every finite group whose order is divisible by fewer than three distinct primes is solvable. In 1897 Burnside's classic work Theory of Groups of Finite Order was published. The second edition (published 1911) was for many decades the standard work in the field. A major difference between the editions was the inclusion of character theory in the second. Burnside is also remembered for the formulation of Burnside's problem that concerns the question of bounding the size of a group if there are fixed bounds both on the order of all of its elements and the number of elements needed to generate it, and also for Burnside's lemma (a formula relating the number of orbits of a permutation group acting on a set with the number of fixed points of each of its elements) though the latter had been discovered earlier and independently by Frobenius and Augustin Cauchy. He received an honorary doctorate (D.Sc.) from the University of Dublin in June 1901. In addition to his mathematical work, Burnside was a noted rower. While he was a lecturer at Cambridge, he also coached the rowing crew team. In fact, his obituary in The Times took more interest in his athletic career, calling him 'one of the best known Cambridge athletes of his day'. He is buried at the West Wickham Parish Church in South London.

of T, then $\{x_1, x_2\}$, $\{x_3, x_4\}$ and $\{x_5, x_6\}$ are conjugacy classes of G, and based on proposition 2.1.7 the tuple $(x_1, \ldots, x_6)$ is end-commutable.
We deduce that $2^3 \cdot 3! = 48$ end-commutable orderings of T exist, and based on proposition 2.1.7 we derive $\mid EA(G) \mid = 48 \cdot 8 \cdot 7 = 2688$.
$Aut(G) \cong_{\mathfrak{g}} S_4$ resp. $Inn(G) \cong_{\mathfrak{g}} V_4$ are valid, and therefor proposition 2.1.7 lets us deduce that 112 resp. 672 orbits of $EA(G)$ under the action of $Aut(G)$ resp. of $Inn(G)$ are existing.◇[2]

---

[2]Christian Felix Klein (25 April 1849 to 22 June 1925) was a German mathematician and mathematics educator, known for his work with group theory, complex analysis, non-Euclidean geometry, and on the associations between geometry and group theory. His 1872 Erlangen Program, classifying geometries by their basic symmetry groups, was an influential synthesis of much of the mathematics of the time. Klein's dissertation, on line geometry and its applications to mechanics, classified second degree line complexes using Weierstrass's theory of elementary divisors. Klein's first important mathematical discoveries were made during 1870. In collaboration with Sophus Lie, he discovered the fundamental properties of the asymptotic lines on the Kummer surface. They later investigated W-curves, curves invariant under a group of projective transformations. It was Lie who introduced Klein to the concept of group, which was to have a major role in his later work. Klein also learned about groups from Camille Jordan. Felix Klein was born on 25 April 1849 in Düsseldorf; his father, Caspar Klein (1809 to 1889), was a Prussian government official's secretary stationed in the Rhine Province. Klein's mother was Sophie Elise Klein (1819 to 1890, née Kayser). He attended the Gymnasium in Düsseldorf, then studied mathematics and physics at the University of Bonn, 1865 to 1866, intending to become a physicist. At that time, Julius Plücker had Bonn's professorship of mathematics and experimental physics, but by the time Klein became his assistant, during 1866, Plücker's interest was geometry. Klein received his doctorate, supervised by Plücker, from the University of Bonn during 1868. Plücker died during 1868, leaving his book concerning the basis of line geometry incomplete. Klein was the obvious person to complete the second part of Plücker's Neue Geometrie des Raumes, and thus became acquainted with Alfred Clebsch, who had relocated to Göttingen during 1868. Klein visited Clebsch the next year, along with visits to Berlin and Paris. During July 1870, at the beginning of the Franco-Prussian War, he was in Paris and had to leave the country. For a brief time he served as a medical orderly in the Prussian army before being appointed lecturer at Göttingen during early 1871. Erlangen appointed Klein professor during 1872, when he was only 23 years old. For this, he was endorsed by Clebsch, who regarded him as likely to become the best mathematician of his time. Klein did not desire a school at Erlangen where there were few students, and so he was pleased to be offered a professorship at Munich's Technische Hochschule during 1875. There he and Alexander von Brill taught advanced courses to many excellent students, including, Adolf Hurwitz, Walther von Dyck, Karl Rohn, Carl Runge, Max Planck, Luigi Bianchi, and Gregorio Ricci-Curbastro. During 1875 Klein married Anne Hegel, the granddaughter of the philosopher Georg Wilhelm Friedrich Hegel. After five years at the Technische Hochschule, Klein was appointed to a chair of geometry at Leipzig. There his colleagues included Walther von Dyck, Rohn, Eduard Study and Friedrich Engel. Klein's years at Leipzig, 1880 to 1886, fundamentally changed his life. During 1882, his health collapsed; during 1883 to 1884, he was plagued by depression. Nonetheless his research continued; his seminal work on hyperelliptic sigma functions dates from around this period, being published during 1886 and 1888. Klein accepted a professorship at the University of Göttingen during 1886. From then until his 1913 retirement, he sought to re-establish Göttingen as the world's main mathematics research center. Yet he never managed to transfer from Leipzig to Göttingen his own primacy as a developer of geometry. At Göttingen, he taught a variety of courses, mainly

concerning the interface between mathematics and physics, such as mechanics and potential theory. The research facility Klein established at Göttingen served as a model for the best such facilities throughout the world. He introduced weekly discussion meetings, and created a mathematical reading room and library. During 1895, Klein hired David Hilbert away from the University of Königsberg; this appointment proved fateful, because Hilbert continued Göttingen's good reputation until his own retirement during 1932. With Klein's editorship, Mathematische Annalen became one of the best mathematics journals in the world. Founded by Clebsch, only with Klein's management did it first rival then surpass Crelle's Journal based in the University of Berlin. Klein established a small team of editors who met regularly, making democratic decisions. The journal specialized in complex analysis, algebraic geometry, and invariant theory (at least until Hilbert ended the subject). It also provided an important outlet for real analysis and the new group theory. During 1893 in Chicago, Klein was a major speaker at the International Mathematical Congress held as part of the World's Columbian Exposition. Due partly to Klein's efforts, Göttingen began admitting women during 1893. He supervised the first Ph.D. thesis in mathematics written at Göttingen by a woman; she was Grace Chisholm Young, an English student of Arthur Cayley's, whom Klein admired. During 1897 Klein became a foreign member of the Royal Netherlands Academy of Arts and Sciences. About 1900, Klein began to become interested in mathematical instruction in schools. During 1905, he was decisive in formulating a plan recommending that analytic geometry, the rudiments of differential and integral calculus, and the function concept be taught in secondary schools. This recommendation was gradually implemented in many countries around the world. During 1908, Klein was elected president of the International Commission on Mathematical Instruction at the Rome International Congress of Mathematicians. With his guidance, the German part of the Commission published many volumes on the teaching of mathematics at all levels in Germany. The London Mathematical Society awarded Klein its De Morgan Medal during 1893. He was elected a member of the Royal Society during 1885, and was awarded its Copley Medal during 1912. He retired the next year due to ill health, but continued to teach mathematics at his home for some years more. Klein had the title of Geheimrat (trusted-advisor). He died in Göttingen during 1925. Klein devised the 'Klein bottle' named after him, a one-sided closed surface which cannot be embedded in three-dimensional Euclidean space, but it may be immersed as a cylinder looped back through itself to join with its other end from the 'inside'. It may be embedded in the Euclidean space of dimensions 4 and higher. The concept of a Klein Bottle was devised as a 3-Dimensional Möbius strip, with one method of construction being the attachment of the edges of two Möbius strips. During the 1890s, Klein began studying mathematical physics more intensively, writing on the gyroscope with Arnold Sommerfeld. During 1894, he initiated the idea of an encyclopedia of mathematics including its applications, which became the Enzyklopädie der mathematischen Wissenschaften. This enterprise, which endured until 1935, provided an important standard reference of enduring value. During 1871, while at Göttingen, Klein made major discoveries in geometry. He published two papers On the So-called Non-Euclidean Geometry showing that Euclidean and non-Euclidean geometries could be considered metric spaces determined by a Cayley-Klein metric. This insight had the corollary that non-Euclidean geometry was consistent if and only if Euclidean geometry was, giving the same status to geometries Euclidean and non-Euclidean, and ending all controversy about non-Euclidean geometry. Arthur Cayley never accepted Klein's argument, believing it to be circular. Klein's synthesis of geometry as the study of the properties of a space that is invariant under a given group of transformations, known as the Erlangen Program (1872), profoundly influenced the evolution of mathematics. This program was initiated by Klein's inaugural lecture as professor at Erlangen, although it was not the actual speech he gave on the occasion. The program proposed a unified system of geometry that has become the accepted modern method.

Within the next section we present a construction of end-commutable orderings for a nilpotent group and its conjugacy classes.

2.2 End-commutable orderings of conjugacy classes

2.2.1 Proposition

Let G be a finite nilpotent group, C a conjugacy class of G and T a subset of C. If G is $\mathcal{G}$-generated by T, then G is a cyclic group.

Proof. Two conjugate elements of G are identical modulo G'. Based on the assumption we deduce that G/G' is cyclic. By using a theorem of Wielandt[3]

---

[3] Klein showed how the essential properties of a given geometry could be represented by the group of transformations that preserve those properties. Thus the program's definition of geometry encompassed both Euclidean and non-Euclidean geometry. Presently the significance of Klein's contributions to geometry is more than evident, but not because those contributions are now considered strange or wrong. On the contrary, those contributions have become so much a part of our present mathematical thinking that it is difficult for us to appreciate their novelty, and the way in which they were not immediately accepted by all his contemporaries. Klein saw his work on complex analysis as his major contribution to mathematics, specifically his work on the link between certain ideas of Riemann's and invariant theory, number theory and abstract algebra, group theory and geometry with more than 3 dimensions and differential equations, especially equations he invented, namely elliptic modular functions and automorphic functions. Klein showed that the modular group moves the fundamental region of the complex plane so as to tessellate that plane. During 1879, he examined the action of $PSL(2,7)$, considered as an image of the modular group, and obtained an explicit representation of a Riemann surface now termed the Klein quartic. He showed that surface was a curve in projective space, that its equation was $x^3y + y^3z + z^3x = 0$, and that its group of symmetries was $PSL(2,7)$ of order 168. His Ueber Riemann's Theorie der algebraischen Funktionen und ihre Integrale (1882) treats complex analysis in a geometric way, connecting potential theory and conformal mappings. This work drew on notions from fluid dynamics. Klein considered equations of degree > 4, and was especially interested in using transcendental methods to solve the general equation of the fifth degree. Building on the methods of Charles Hermite and Leopold Kronecker, he produced similar results to those of Brioschi and later completely solved the problem by means of the icosahedral group. This work enabled him to write a series of papers on elliptic modular functions. In his 1884 book on the icosahedron, Klein established a theory of automorphic functions, associating algebra and geometry. However Poincaré published an outline of his theory of automorphic functions during 1881, which resulted in a friendly rivalry between the two men. Both sought to state and prove a grand uniformization theorem that would establish the new theory more completely. Klein succeeded in formulating such a theorem and in describing a strategy for proving it. But while doing this work his health decreased, as mentioned above. Klein summarized his work on automorphic and elliptic modular functions in a four volume treatise, written with Robert Fricke during a period of about 20 years.

[3] Helmut Wielandt (19 December 1910 Niedereggenen, Lörrach, Germany to 14 February 2001) was a German mathematician who worked on permutation groups. He gave a plenary lecture Entwicklungslinien in der Strukturtheorie der endlichen Gruppen (Lines of Development in the Structure Theory of Finite Groups) at the ICM in 1958 at Edinburgh and was an Invited Speaker with talk Bedingungen für die Konjugiertheit von Untergrup-

End-commutable orderings and exponents 45

for nilpotent groups we conclude $G' \leq \Phi(G)$. Thus, the Frattini factor group of G is cyclic, and we derive that G is cyclic because the elements of the Frattini subgroup can be omitted from every generating set of G.◊

2.2.2 Lemma (end-commutable orderings for conjugacy classes)

Let G be a finite nilpotent group. Every conjugacy class of G is end-commutable.

Proof. We prove the lemma by using an induction argument based on the order of the group. If G is Abelian, then the statement is straightforward to prove. Let G be a non-Abelian group and C a non-central conjugacy class of G. Based on proposition 2.2.1 we can assume that $N := \langle C \rangle_{\mathcal{G}}$ is a proper normal subgroup of G containing C. The conjugacy class C of G decomposes in N into conjugacy classes of N (because they are invariant under N). Let $n \in \mathbb{N}$ and $C_1, \ldots, C_n$ conjugacy classes of N such that C is the disjoint union of the sets $C_1, \ldots, C_n$. N is a proper subgroup of G and thus we can use an induction argument and deduce that for all $i \in \underline{n}$ an end-commutable ordering $(c_{i,1}, \ldots, c_{i,r_i})$ of C_i exists. For all $i \in \underline{n}$ the element $\overline{C_i}$ is a central element of KN (K an arbitrary field), and thus we us part (iii) of proposition 2.1.7 to obtain the end-commutable ordering $(c_{1,1}, \ldots, c_{1,r_1}, \ldots, c_{n,1}, \ldots, c_{n,r_n})$ of C by concatenating the end-commutable orderings for all C_i.◊

2.2.3 Corollary (end-commutable orderings for a nilpotent group)

If G is a finite nilpotent group, then G is end-commutable.

Proof. Based on lemma 2.2.2 every conjugacy class of G is end-commutable. If C is a conjugacy class of of G, then $\overline{C}$ is central in KG (K an arbitrary field). The proof is a consequence of part (iii) of proposition 2.1.7.◊

2.2.4 A construction method

Let G be a finite nilpotent group. The induction argument within lemma 2.2.2 is used to derive a construction method for an end-commutable ordering for a conjugacy class C of G:
within the normal subgroup $N := \langle C \rangle_{\mathcal{G}}$ the class C decomposes into conjugacy classes $C_1, \ldots, C_n$ of N. If we have constructed an end-commutable ordering for all of these classes, then we use part (iii) of proposition 2.1.7 to obtain an end-commutable ordering for C by concatenating these orderings.

---

pen endlicher Gruppen (Conditions for the Conjugacy of Finite Groups) at the ICM in 1962 in Stockholm.

For the classes $C_1, \ldots, C_n$ of N the same argumentation is valid as for the class C of G. This decomposition process has to be done as long as we obtain classes which consists of pairwise commuting elements. Proposition 2.2.1 guaranties this approach.

If we have determined for every conjugacy class of G an end-commutable ordering, then we can use corollary 2.2.3 and part (iii) of proposition 2.1.7 to obtain an end-commutable ordering of G by concatenating the orderings. This method is presented within the next two examples:

(i) Let $G := D_{16}$, $a, b \in G$, $G = \langle a, b \rangle_{\mathfrak{g}}$, $o(a) = 8$, $o(b) = 2$ and $a^b = a^{-1}$. $Z(G) = \{1_G, a^4\}$ is valid, and $C_1 := a^G = \{a, a^7\}$, $C_2 := (a^2)^G = \{a^2, a^6\}$, $C_3 := (a^3)^G = \{a^3, a^5\}$, $C_4 := (ab)^G = \{ab, a^3b, a^5b, a^7b\}$ as well as $C_5 := b^G = \{b, a^2b, a^4b, a^6b\}$ are the non-central conjugacy classes of G. The elements within C_1, C_2 and C_3 are commuting.

Because of $(ab)^{(a^2b)} = a^5b$, $(a^3b)^{(ab)} = a^7b$ and proposition 2.2.1 the class C_4 decomposes in $\langle C_4 \rangle_{\mathfrak{g}}$ into the classes $\{ab, a^5b\}$ and $\{a^3b, a^7b\}$ of $\langle C_4 \rangle_{\mathfrak{g}}$ which consist of pairwise commuting elements. Hence, (ab, a^5b, a^3b, a^7b) is an end-commutable ordering of C_4. This one is presented within the example 2.1.1. Because of $(a^2b)^b = a^6b$, $(a^4b)^{(a^2b)} = b$ and proposition 2.2.1 we deduce that C_5 decomposes in $\langle C_5 \rangle_{\mathfrak{g}}$ into the two classes $\{a^2b, a^6b\}$ and $\{b, a^4b\}$ of $\langle C_5 \rangle_{\mathfrak{g}}$ which consist of pairwise commuting elements. Therefor, (a^2b, a^6b, b, a^4b) is an end-commutable ordering of C_5. We deduce that the tuple

$$(1_G, a^4, a, a^7, a^2, a^6, a^3, a^5, ab, a^5b, a^3b, a^7b, a^2b, a^6b, a^4b, b)$$

is an end-commutable ordering of G.

For the second example the next graphic is useful:

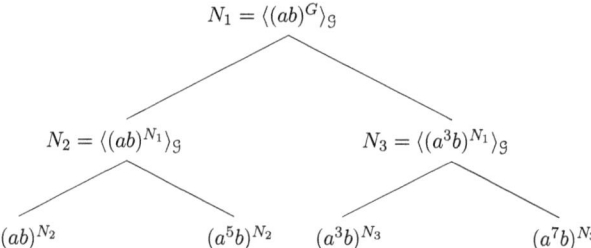

(ii) Let $G := D_{32}$, $a, b \in G$, $G = \langle a, b \rangle_{\mathfrak{g}}$, $o(a) = 16$, $o(b) = 2$ and $a^b = a^{-1}$. The set $(ab)^G = \{ab, a^3b, a^5b, a^7b, a^9b, a^{11}b, a^{13}b, a^{15}b\}$ is a conjugacy class of G, and it is decomposing in $\langle (ab)^G \rangle_{\mathfrak{g}}$ into the two conjugacy classes $\{ab, a^5b, a^9b, a^{13}b\}$ and $\{a^3b, a^7b, a^{11}b, a^{15}b\}$ of $\langle (ab)^G \rangle_{\mathfrak{g}}$.
The first class decomposes in $\langle \{ab, a^5b, a^9b, a^{13}b\} \rangle_{\mathfrak{g}}$ into the classes $\{ab, a^9b\}$

End-commutable orderings and exponents 47

and $\{a^5b, a^{13}b\}$, the second class in $\langle\{a^3b, a^7b, a^{11}b, a^{15}b\}\rangle_{\mathfrak{G}}$ into the classes $\{a^3b, a^{11}b\}$ and $\{a^7b, a^{15}b\}$. All four sets of order 2 are possessing pairwise commuting elements. Hence, $(ab, a^9b, a^5b, a^{13}b, a^3b, a^{11}b, a^7b, a^{15}b)$ is an end-commutable ordering of $(ab)^G$ in KG. ⋄

2.3 End-commutable orderings and nilpotent finite groups

2.3.1 Example

In cycle notation the conjugacy classes of S_3 are exactly $C_1 := \{(1)\}$, $C_2 := \{(12),(13),(23)\}$ and $C_3 := \{(123),(132)\}$. C_1 and C_3 consist of pairwise commuting elements, and thus they are end-commutable. All elements of the set C_2 are pairwise non-commuting, and therefor this set is not end-commutable (because the last two elements of an end-commutable ordering are commuting). The next lemma lets us deduce that G is not-end-commutable, too. ⋄

2.3.2 Lemma

Let G be a finite group. The following statements are equivalent:

(i) G is end-commutable.

(ii) Every conjugacy class of G is end-commutable.

Proof. The implication from (ii) to (i) is deductable from part (iii) of proposition 2.1.7. For the opposite implication, let G be end-commutable by $Q := (g_1, \ldots, g_r)$ and C be a conjugacy class of G. We define recursively: let i be minimal within $\underline{r}$ such that $g_i \in C$ is valid. We define $a_1 := g_i$. If a_j is defined, then we choose k minimal within the set $M := \underline{r} \setminus \{t \mid \exists l \in \underline{j} : g_t = a_l\}$ such that $g_k \in C$ is valid and M is non-empty. In this case we define $a_{j+1} := g_k$. If M is empty, then the definition is complete. By definition $C = \{a_1, \ldots, a_{|C|}\}$ is true, and we prove that $Q_C := (a_1, \ldots, a_{|C|})$ is an end-commutable ordering of C. If $i \in \lfloor |C| \rfloor$ is true, then an element $t \in \underline{r}$ exists such that $a_i = g_t$ is valid. Let $X := \overline{\{g_{t+1}, \ldots, g_r\}} \setminus \{a_{i+1}, \ldots, a_{|C|}\}$. Q is an end-commutable ordering of G, and thus $\{a_{i+1}, \ldots, a_{|C|}\}^{a_i} \cup X^{a_i}$ $= \{a_{i+1}, \ldots, a_{|C|}\} \cup X$ is true. If $j \in \{i+1, \ldots, \lfloor |C| \rfloor\}$, then $a_j^{a_i} \in \{a_{i+1}, \ldots, a_{|C|}\} \cup X$ is valid. By definition the set X contains no element of C. We use $a_j^{a_i} \in C$ and deduce $a_j^{a_i} \in \{a_{i+1}, \ldots, a_{|C|}\}$. Thus, $\{a_{i+1}, \ldots, a_{|C|}\}^{a_i} = \{a_{i+1}, \ldots, a_{|C|}\}$ is true. Therefor, Q_C is an end-commutable ordering of C. ⋄

2.3.3 Corollary

Let G be a finite group. G is end-commutable if and only if every normal subset of G is end-commutable.

Proof. Based on lemma 2.3.2 we deduce that with G also every conjugacy class of G is end-commutable. If T is a normal subset of G, then T is the disjoint union of conjugacy classes of G. We use part (iii) of proposition 2.1.7 to finalize the proof.◇

2.3.4 Lemma

Let G be a finite end-commutable group. If every proper normal subgroup of G is nilpotent, then G is nilpotent, too.

Proof. Let us assume that G is not nilpotent. By using the assumption every proper normal subgroup of G is contained in the Fitting subgroup $F(G)$ of G. Let $x \in G \setminus F(G)$. $N := \langle x^G \rangle_{\mathfrak{g}}$ is a non trivial normal subgroup of G. If $N \neq G$ would be valid, then N would be contained in $F(G)$ contradicting $x \notin F(G)$. Hence, $N = G$ is true. Based on lemma 2.3.2 the conjugacy class x^G possesses an end-commutable ordering $(g_1, \ldots, g_n)$. For all $i \in \underline{n}$ we define $U_i := \langle g_i, \ldots, g_n \rangle_{\mathfrak{g}}$. By definition of the end-commutable ordering for all $i \in \underline{n} \setminus \underline{1}$ the subgroup U_i is a normal subgroup of U_{i-1}. Because of $x \notin F(G)$ none of the subgroups $U_1, \ldots, U_n$ is contained in $F(G)$. Hence, $G = U_n = \langle g_n \rangle_{\mathfrak{g}}$ is valid.◇

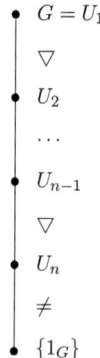

As used within lemma 2.3.4 we have proven that a finite end-commutable group G possesses a chain of subnormal subgroups of length $|G|$. This is also valid for nilpotent groups which is a consequence of main theorem 2.3.6 and will be proven after the next preliminary lemma.

End-commutable orderings and exponents 49

2.3.5 Lemma

If G is an end-commutable finite group, then G is nilpotent.

Proof. We prove this lemma by an induction argument based on the order of G. If G is trivial, then G is nilpotent. Let N be a proper normal subgroup of G. N is a normal subset of G and possesses based on corollary 2.3.3 an end-commutable ordering. We use an induction argument to deduce that every proper normal subgroup of G is nilpotent. By applying lemma 2.3.4 we finalize the proof. ◇

The following main theorem presents a nilpotency criteria for finite groups based on end-commutable orderings:

2.3.6 Main theorem (nilpotency criteria and end-commutable-orderings)

Let G be a finite group. The following statements are equivalent:

(i) G is nilpotent.

(ii) G is end-commutable.

(iii) Every conjugacy class of G is end-commutable.

(iv) Every normal subset of G is end-commutable.

(v) G possesses exactly one maximal end-commutable subset.

Proof. The statements (ii), (iii) and (iv) are equivalent based on lemma 2.3.2 and corollary 2.3.3. The implication from (i) to (ii) is the content of corollary 2.2.3, and the implication from (ii) to (i) is proven within lemma 2.3.5. The implication from (ii) to (v) is true because G is the only maximal end-commutable subset of G. If only one maximal end-commutable subset of G exists, then this subset is G because for every element $g \in G$ the set $\{g\}$ is end-commutable. ◇

One application of the main theorem is the following enhancement of proposition 2.2.1:

2.3.7 Corollary (nilpotency criteria and cyclic subgroups)

Let G be a finite group. The following statements are equivalent:

(i) G is nilpotent.

(ii) Every subgroup U of G generated by a conjugacy class of G is cyclic.

(iii) Every subgroup U of G generated by a subset of a conjugacy class of G is cyclic.

Proof. The implication from (i) to (iii) is true based on proposition 2.2.1, and the implication from (iii) to (ii) is straightforward to verify.
We use an induction argument based on the order of G to prove the implication from (ii) to (i). We only have to deduce based on theorem 2.3.6 that every conjugacy class of G is end-commutable. Let $g \in G \setminus Z(G)$ and $N := \langle g^G \rangle_\mathfrak{g}$. If $N = G$ is true, then G is cyclic by our assumption. Let N be a proper normal subgroup of G. The conjugacy class g^G decomposes in N in conjugacy classes $C_1, \ldots, C_n$ of N. The precondition of the induction argument is compatible with subgroups of G, and thus for every conjugacy class $C_1, \ldots, C_n$ an end-commutable ordering exist. We use part (iii) of proposition 2.1.7 to construct an end-commutable ordering of g^G. By using main theorem 2.3.6 the proof is finished.◇

2.3.8 Corollary

If G is a finite nilpotent group, then every non-central conjugacy class of G possess two commuting elements.

Proof. Let C be a non-central conjugacy class of G. Based on theorem 2.3.6 the class C possess an end-commutable ordering $(c_1, \ldots, c_n)$, and by definition the elements c_{n-1} and c_n are commuting. ◇

Corollary 2.3.8 can be proven alternatively by an induction argument on the order of G. We can enhance the corollary to arbitrary finite groups in the following way:

2.3.9 Proposition

Every non-Abelian finite group possesses at least one conjugacy class containing two commuting elements.

Proof. Let G be a finite group. We prove the proposition by an induction argument based on order of G. The beginning of the induction is straightforward to prove, and based on corollary 2.3.8 we can assume that G is not nilpotent. If a non-Abelian proper subgroup of G would exist, then two conjugate and commuting elements would exist within this subgroup. Thus, also G would possess these elements and the proof would be finished. Therefor we assume that every proper subgroup of G is Abelian. In particular, G is minimal non-nilpotent. We use the theorem on page 181 in [26] and conclude that $|G|$ possess two distinct prime divisors p and q and an Abelian normal p-Sylow subgroup P and a cyclic q-Sylow subgroup Q such that (P, Q) is a semi direct decomposition of G. G is non-Abelian, and thus

End-commutable orderings and exponents 51

elements $g \in P$ and $h \in Q$ exist such that $g \neq g^h$ is true. P is a normal and Abelian subgroup of G. Therefor the conjugate elements g and g^h are commuting. ⋄

2.3.10 Remark

Let G, H be finite groups, U a subgroup and N a normal subgroup of G. If G and H are possessing end-commutable orderings $Q_G := (g_1, \ldots, g_n)$ and $Q_H := (h_1, \ldots, h_r)$, then G and H are nilpotent based on theorem 2.3.6. Thus, U, G/N and $G \times H$ are nilpotent, too. Again – by using theorem 2.3.6 – these groups are possessing end-commutable orderings, and we want to construct these by using Q_G and Q_H.

The method used within lemma 2.3.2 is usable for U and G/N: delete are entries of Q_G which are not contained in U. The resulting tuple is an end-commutable ordering of U. Calculate all entries in Q_G modulo N and delete – beginning at $g_n N$ – all duplicates then the resulting tuple is an end-commutable ordering of G/N. Part (iv) of proposition 2.1.7 can be used to deduce that $(g_1 h_1, \ldots, g_1 h_r, \ldots, g_n h_1, \ldots, g_n h_r)$ is an end-commutable ordering of $G \times H$. ⋄

2.3.11 Remark

Let G be a finite group and T a subset of G.

(i) **Polycyclic groups:** Based on theorem 2.3.6 we know that G is nilpotent if and only if G possesses an end-commutable ordering $(g_1, \ldots, g_n)$. Remark 2.1.3 lets us deduce that for all $i \in \underline{n-1}$, the sets $\{(g_{i+1})^{g_i}, \ldots, (g_n)^{g_i}\}$ and $\{g_{i+1}, \ldots, g_n\}$ are identical: g_i is normalizing the set $\{g_{i+1}, \ldots, g_n\}$. Thus, the subgroup $\langle g_{i+1}, \ldots, g_n \rangle_\mathcal{G}$ is normal in $\langle g_i, \ldots, g_n \rangle_\mathcal{G}$ possessing a cyclic factor group. By using this procedure we can construct a subnormal series with cyclic factors from 1 to G. Such groups are called polycyclic. We have proven the well-known theorem that finite nilpotent groups are polycyclic but in a very constructive way. In addition, it is well-known that the finite polycyclic groups are exactly the finite solvable groups.

(ii) **Maximal end-commutable sets:** By using a similar construction we can prove that an end-commutable subset T leads to a solvable subgroup $\langle T \rangle_\mathcal{G}$. This span need not to be nilpotent. For this, let us focus on the symmetric group S_3 containing an element a of order 3 and three involutions b, c, d. The set $\{1, a, a^2\}$ is a normal subgroup ($= A_3$) of S_3 consisting of commutable elements. If we take the involution b, then $(b, a^2, a, 1)$ is an end-commutable ordering of $M_b := \{1, a, a^2, b\}$. The subgroup generated by M_b is S_3 which is solvable but not nilpotent. This examples also shows us that maximal end-commutable sets are not subgroups in general: the set M_b

is maximal end-commutable or is contained in a maximal end-commutable set of order 5 because S_3 is not end-commutable.

(iii) **Maximal end-commutable orderings:** In part (ii) maximal end-commutable sets are regarded. Of course, every end-commutable set can be enhanced to a maximal end-commutable set. Let S be an end-commutable set with end-commutable ordering $(s_1 \ldots, s_n)$ where $n := \mid S \mid$. In a maximal end-commutable ordering $S \subseteq M$ the corresponding end-commutable ordering of M might not be related to the one of S. It is unclear if the ordering of S can be enhanced to the right or left to an end-commutable ordering of M. Enhancing to the right would imply that the new item is centralizing all previous ones which should be relatively rare. Enhancing to the left is possible if the new item is normalizing the set of all previous ones which might be more easier to achieve. An interesting topic is the connection between maximal end-commutable sets and orderings and the enhancements of arbitrary end-commutable sets and orderings.

If we take as S for example a Carter subgroup C in a finite solvable group , then C is nilpotent and self-normalizing. C possesses an end-commutable ordering which cannot be enhanced to the left or right because C is containing its own normalizer. So Carter subgroups are exactly those subgroups which possess an end-commutable ordering not enlargeable to the left.

All of these topics are not fully understood by the author and remains as an open topic.◇

2.4 The exponent of the center

2.4.1 Definition and remark (monotone bijection)

Let T be a finite subset of $\mathbb{N}$. With respect to the natural order on $\lfloor T \rfloor$ and T exactly one monotone bijection between $\lfloor T \rfloor$ and T exists which we denote by φ_T.◇

2.4.2 Definition (subsets of a given order)

Let T be a set and $i \in \mathbb{N}_0$. By $\binom{T}{i}$ we denote the set of subsets of T of order i.◇

The following proposition is straightforward to be proven by an induction argument:

2.4.3 Proposition

Let A be an associative K-algebra, $n \in \mathbb{N}$ and $x_1, \ldots, x_n \in A$. The identity
$$x_1 * \cdots * x_n = \sum_{i=1}^{n} \sum_{T \in \binom{n_i}{i}} x_{(1\varphi_T)} \cdots x_{(i\varphi_T)}$$
is valid. ⋄

2.4.4 Corollary

Let A be an associative K-algebra, $n \in \mathbb{N}$ and $a \in A$. The following statements are valid:

(i) $\underbrace{a * \cdots * a}_{n-times} = \sum_{i=1}^{n} \binom{n}{i}_K a^i$

(ii) If p is a prime number and $char(K) = p$, then $\underbrace{a * \cdots * a}_{p^n-times} = a^{(p^n)}$ is valid.

Proof. ad(i): Based on proposition 2.4.3 we deduce
$$\underbrace{a * \cdots * a}_{n-times} = \sum_{i=1}^{n} \sum_{T \in \binom{n_i}{i}} a^i = \sum_{i=1}^{n} | \binom{n_i}{i} |_K a^i = \sum_{i=1}^{n} \binom{n}{i}_K a^i.$$

ad(ii): For all $i \in \overline{p^n - 1}$, it is well-known that the prime number $p = char(K)$ is a divisor of $\binom{p^n}{i}$, and thus (ii) is a consequence of (i). ⋄

2.4.5 Proposition

Let p be a prime number, G a finite p-group, K a field and $char(K) = p$. If G is Abelian, then for all $n \in \mathbb{N}$ the identity
$$(1_G + rad(KG))^{p^n} = 1_G + rad(K^{p^n} G^{p^n})$$
is true. In particular, $1_G + rad(KG)$ is a torsion group and
$$exp(G) = exp(1_G + rad(KG))$$
is valid.

Proof. Let G be Abelian. KG is commutative, and we use $char(K) = p$ and the binomial theorem to prove for all $a, b \in KG$ the identity

(1) $(a + b)^p = a^p + b^p$.

Based on theorem 1.1.21 we deduce $rad(KG) = Aug(KG)$. If $x \in rad(KG)$ is valid, then for every $g \in G \setminus \{1_G\}$ an element $k_g \in K$ exists such that $x = \sum_{g \in G \setminus \{1\}} k_g(g - 1_G)$ is true. If $n \in \mathbb{N}$, then we deduce by an iterative

argumentation of (1) that $(1_G+x)^{p^n} = 1_G + \sum_{g \in G \setminus \{1_G\}} k_g^{p^n}(g^{p^n} - 1_G)$ is valid.
This equation and theorem 1.1.21 finalize the proof.◇

2.4.6 Remark

Let G be a group, $n \in \mathbb{N}$, $a \in G$ and $b \in a^G$. If a^n is central in G, then $a^n = b^n$ is true.◇

If A is an associative K-algebra such that every element of A is star regular, then we use the symbol A^* instead of $Q(A)$. The following proposition reduces the determination of the exponent of $Z(rad(KG)^*)$ to a sub-problem:

2.4.7 Proposition (exponent of the center)

Let p be a prime number, K a field, $char(K) = p$ and G a finite p-group. The following statements are valid:

(i) $Z(rad(KG)^*)$ is a torsion group.

(ii) The order of each element of $Z(rad(KG)^*)$ is a p-power.

(iii) $exp(Z(rad(KG)^*)) = max\{exp(Z(G)), max\{o(\overline{g^G}) \mid g \in G \setminus Z(G)\}\}$

Proof. Based on corollary 2.4.4 for all $x \in KG$ and for all $n \in \mathbb{N}$ the identity $\underbrace{x * \cdots * x}_{p^n-times} = x^{(p^n)}$ is valid. Let $g \in G \setminus Z(G)$. G is a nilpotent group, and thus we deduce from theorem 2.3.6 that g^G possesses an end-commutable ordering $(a_1, \ldots, a_r)$. We use theorem 2.1.5 to conclude for all $n \in \mathbb{N}$ the identity $(\overline{g^G})^{(p^n)} = \sum_{i=1}^{r} a_i^{(p^n)}$. Hence, based on remark 2.4.6 the equation $(\overline{g^G})^{exp(G/Z(G))} = \sum_{i=1}^{r} a_i^{exp(G/Z(G))} = \sum_{i=1}^{r} g^{exp(G/Z(G))} = 0_{KG}$ is true. Thus, we have proven the parts (i) and (ii). In addition, proposition 2.4.5 lets us deduce that for all $r \in rad(KZ(G))$ the identity $r^{exp(Z(G))} = 0_{KG}$ is true. If $x \in Z(rad(KG))$, then based on proposition 1.3.11 elements $r \in rad(KZ(G))$, $n \in \mathbb{N}$, $g_1, \ldots, g_n \in G \setminus Z(G)$ and $k_1, \ldots, k_n \in K$ exist such that $x = r + \sum_{i=1}^{n} k_i \overline{(g_i^G)}$ is valid. Because of the commutativity of $Z(rad(KG))$ and the binomial theorem as well as corollary 2.4.4 for all $e \in \mathbb{N}$ the statement $\underbrace{x * \cdots * x}_{p^e-times} = x^{(p^e)} = r^{(p^e)} + \sum_{i=1}^{n} k_i^{(p^e)} (\overline{g_i^G})^{(p^e)}$ is true. We use this identity and (i) to finish the proof. ◇

This proposition reduces the determination of the exponent of $Z(rad(KG)^*)$ to the calculation of the orders of the conjugacy class sums. These orders can

End-commutable orderings and exponents   55

be determined based on our results of end-commutable orderings. Another approach to the parts (ii) to (iv) of the following theorem can be found in the article [9] of A.A. Bovdi and Z. Patay. In particular, part (iv) demonstrates us how the exponent of $Z(rad(KG)^*)$ can be determined purely within the group G. The field K has no effect on this calculation (But the field has an effect on the invariants of the center which is presented later in chapter 4.). Part (i) generalizes the theorem of A.A. Bovdi and Z. Patay to nilpotent groups. Part (v) shows us that the concept of end-commutable orderings is useful also for other topics.

2.4.8 Theorem (order of class sums)

Let p be a prime number, G a finite nilpotent group, K a field, $char(K) = p$ and $g \in G \setminus Z(G)$. The following statements are valid:

(i) $(\overline{g^G})^p = (\frac{|C_G(g^p)|}{|C_G(g)|})_K \overline{(g^G)^p}$

(ii) If G is a p-group such that $C_G(g) < C_G(g^p)$ is valid, then $(\overline{g^G})^p = 0_{KG}$ is true.

(iii) If G is a p-group such that $C_G(g) = C_G(g^p)$ is valid, then $(\overline{g^G})^p = \overline{(g^p)^G}$ is true.

(iv) If G is a p-group, then $o(\overline{g^G}) = p^{min\{n \in \mathbb{N} \mid C_G(g) < C_G(g^{p^n})\}}$ is valid. In particular, $\overline{g^G}$ is nilpotent and $\frac{o(\overline{g^G})}{p} < cl(\overline{g^G}) \leq o(\overline{g^G})$ is valid.

(v) If g is a p'-element of G, then $\overline{g^G}$ is not nilpotent.

Proof. ad(i): Based on corollary 2.4.4 we deduce

(1) $\underbrace{\overline{g^G} * \cdots * \overline{g^G}}_{p-times} = (\overline{g^G})^p \in Z(rad(KG))$.

G is a nilpotent group, and thus g^G possesses based on theorem 2.3.6 an end-commutable ordering. We use (1) and theorem 2.1.5 to conclude

(2) $\underbrace{\overline{g^G} * \cdots * \overline{g^G}}_{p-times} = \sum_{x \in g^G} x^p$.

If $x \in g^G$, then x^p and g^p are conjugated. Based on (1), (2) and the fact that the class sums form a basis of the center we deduce that an element $k \in K$ exists such that $(\overline{g^G})^p = k \overline{(g^p)^G}$ is true. The conjugacy class of g resp. of g^p is possessing $\frac{|G|}{|C_G(g)|}$ resp. $\frac{|G|}{|C_G(g^p)|}$ elements. Therefor, $k = (\frac{|C_G(g^p)|}{|C_G(g)|})_K$ is valid, and part (i) is proven.

ad(ii),(iii): G is a p-group and $char(K) = p$ is valid. Hence, (ii) and (iii) are a consequence of (i).

ad(iv): This statement is deductable by using the parts (i), (ii) and (iii).

ad(v): Let g be a p'-element of G. We assume that $\overline{g^G}$ is nilpotent. Let $k \in \mathbb{N}$ such that $cl(\overline{g^G}) \leq p^k$. We deduce that $(\overline{g^G})^{p^k} = 0$ would be valid. Now we use part (i) to deduce that $(\overline{g^G})^{p^k} = \overline{(g^G)^{p^k}}$ is true because g is a p'-element and thus $C_G(g) = C_G(g^p) = \cdots = C_G(g^{p^k})$ is valid. This is a contradiction.⋄

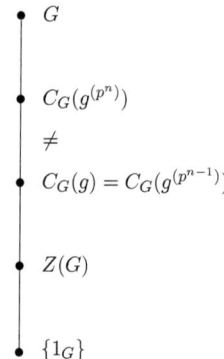

2.4.9 Definition (associated (restricted) Lie algebra)

If A is an associative K-algebra, then we define for all $a, b \in A$

$$a \circ b := ab - ba.$$

$(A; +; \circ)$ is a K-Lie algebra symbolized by A°. It is called the Lie algebra associated to A. If $char(K) = p \in \mathbb{P}$ is valid, then A° is a so-called restricted Lie algebra based on the p-th power map.⋄

Based on theorem 2.4.8 we deduce (because the p-power of a class sum is zero or a class sum of the same length):

2.4.10 Corollary

Let p be a prime number, G a finite p-group, K a field, $char(K) = p$ and $n \in \mathbb{N}$. The K-subspace $\langle \{\overline{g^G} \mid g \in G \setminus Z(G), \mid g^G \mid = p^n\} \rangle_K$ of $Z(rad(KG))$ is a Lie subalgebra of the restricted Lie algebra $(KG)^\circ$.⋄

End-commutable orderings and exponents   57

2.4.11 Example

Let K be a field of order 2, $G := D_{16}$ and $h, a \in G$ such that $G = \langle h, a \rangle_{\mathcal{G}}$, $o(h) = 8$, $o(a) = 2$ and $h^a = h^{-1}$ are valid. If $U := \langle h^2 \rangle_{\mathcal{G}}$, then $a^U = \{a, h^2 a\}$ is true. a and $h^2 a$ are not commuting, and thus a^U does not possess an end-commutable ordering. $(a + h^2 a)^2 = a^2 + a h^2 a + h^2 + (h^2 a)^2 = h^6 + h^2 \neq 0_{KG} = a^2 + (h^2 a)^2$ is true.

This example demonstrates that the exponent of $C_{rad(KG)^*}(U - 1_G)$ (see corollary 1.3.16) might not be analyzable with the concept of end-commutable orderings.⋄

2.5 Bounds

2.5.1 Definition (set of conjugacy classes)

If G is a group, then we use the symbol $\mathcal{K}(G)$ for the set of all conjugacy classes of G.⋄

2.5.2 Proposition

Let p be a prime number, G a finite p-group, K a field and $char(K) = p$. The following statements are valid:

(i) If C is a non-central conjugacy class of G, then

$o(\overline{C}) \mid p^{|\{X \mid X \in \mathcal{K}(G), |X| = |C|\}|}$ is valid. (bound by number of classes of the same length)

(ii) If C is a non-central conjugacy class of G and $c \in C$, then

$o(\overline{C}) \mid o(Z(G)c)$ is valid. (bound by the order in $G/Z(G)$)

(iii) $exp(Z(G)) \mid exp(Z(rad(KG)^*)) \mid max\{exp(Z(G)), exp(G/Z(G))\}$ (bound by $Z(G)$ and $G/Z(G)$)

(iv) $exp(Z(rad(KG)^*)) \mid exp(G)$ (bound by G)

Proof. Parts (i) and (ii) are a consequence of part (iv) of theorem 2.4.8. The other two statements are a consequence of part (ii) and of proposition 2.4.7.⋄

2.5.3 Corollary (min-max bounds)

If p is a prime number, G a non-Abelian p-group, K a field and $char(K) = p$, then $p \leq exp(Z(rad(KG)^*)) \leq \frac{|G|}{p^2}$ is valid.

Proof. G is non-Abelian, and thus $p \mid exp(Z(G))$ is valid. If the factor group by the center is cyclic, then G is Abelian. Therefor, $Z(G)$ possesses an order less or equal to $\frac{|G|}{p^2}$. Hence, $exp(Z(G)) \mid \frac{|G|}{p^2}$ is true.
$Z(G)$ is non-trivial, and therefor $G/Z(G)$ possesses an order less or equal to $\frac{|G|}{p}$. We conclude that the exponent of the non-cyclic group $G/Z(G)$ is less or equal to $\frac{|G|}{p^2}$, and based on part (iii) of proposition 2.5.2 we finish the proof.⋄

2.5.4 Proposition

If p is a prime number, G a finite p-group, U a subgroup of G, K a field, $char(K) = p$ and $u \in U \setminus Z(U)$, then $o(\overline{u^G}) \leq o(\overline{u^U}) \leq o(u)$.

Proof. We start the proof by remarking that u is not central in G. For all $a, b \in U$ such that $C_G(a) = C_G(b)$ is valid an intersection with U yields to the identity $C_U(a) = C_U(b)$. We use theorem 2.4.8 to deduce the first part of the proposition. The second part is deductable by using proposition 2.5.2.⋄

2.5.5 Corollary

Let p be a prime number, G a finite p-group, K a field, $char(K) = p$, $g \in G \setminus Z(G)$ and $r \in \mathbb{N}$ such that $o(\overline{g^G}) = p^r$ is valid. If U is minimal with $g \in U \setminus Z(U)$ and with respect to inclusion, then $\mid U \mid \geq p^{r+2}$ is true. In particular, if $p^r = \frac{|G|}{p^2}$ is true, then G is the only subgroup of G in which g is not central.

Proof. Let us assume that $\mid U \mid < p^{r+2}$ is valid. By using proposition 2.5.4 and corollary 2.5.3 we would deduce $p^r \leq o(\overline{g^U}) \leq \frac{|U|}{p^2} < \frac{p^{r+2}}{p^2} = p^r$ which is a contradiction.⋄

2.5.6 Remark

The minimal possible order p^{r+2} within corollary 2.5.5 is not met in general which is demonstrated at the end of chapter 3 (see Remarks to the bounds within section 2.5).⋄

The following lemma is used later on for analyzing regular p-groups:

2.5.7 Lemma

Let p be a prime number, $n \in \mathbb{N}$, G a non-Abelian p-group, K a field and $char(K) = p$. For all $a, b \in G$ let the element $a^{(p^n)}$ commute with b if and only if $b^{(p^n)}$ commutes with a. Then for every non-central conjugacy class

End-commutable orderings and exponents 59

C of G the statement $o(\overline{C}) \leq p^n$ is valid.

Proof. We prove the lemma by an induction argument based on the order of G. If $\mid G \mid = p^3$ is valid, then corollary 2.5.3 is used for deducting the statement. Let $x \in G \setminus Z(G)$.

<u>Case 1:</u> A maximal subgroup of G containing x exists such that $x \notin Z(U)$ is valid.
Because of $x \in U \setminus Z(U)$ and the compatability of U with induction we deduce by using proposition 2.5.4 the statement $o(\overline{x^G}) \leq o(\overline{x^U}) \leq p^n$. In this case the proof is finished.

<u>Case 2:</u> For every maximal subgroup of G containing x the statement $x \in Z(U)$ is true.
Let U be a maximal subgroup of G containing x. Thus, $x \in Z(U)$ is true. If $y \in G \setminus U$, then we use the nilpotency of G to deduce $G = U\langle y\rangle_{\mathfrak{g}}$. Because of $x \in Z(U) \setminus Z(G)$ the element y is centralized by x. In addition, we use the maximality of U to conclude $y^p \in U$, and thus $[y^p, x] = 1_G$ is valid. In particular, $[x, y^{(p^n)}] = 1_G$ is true, and by using our assumption we derive $[x^{(p^n)}, y] = 1_G$. We have proven that y centralizes the element $x^{(p^n)}$ but not the element x. Theorem 2.4.8 yields to $o(\overline{x^G}) \leq p^n$ and the proof is done.⋄

2.5.8 Remark

Let p be a prime number, G a non-Abelian finite p-group, K a field, $char(K) = p$ and $n \in \mathbb{N}$ such that $p^n = exp(G/Z(G))$ is valid. With this n the assumption of lemma 2.5.7 is fulfilled. If we choose a minimal $n \in \mathbb{N}$ for which these assumption is valid, then we ask whether a non-central conjugacy class of G exists such that their sum is exactly of order p^n. Then the maximal order of all non-central conjugacy class sums of G is exactly p^n. But we will provide a counterexample for this question at the end of chapter 3 (see Remarks to the bounds within section 2.5).⋄

The results so far concerning bounds will be used within chapter 3 to estimate and determine the exponent of the center of the radical. We end this chapter by demonstrating how we can bound the exponent of $Z(rad(KG)^*)$ for a normal subgroup N of G by $exp(Z(rad(KN)^*))$ and $exp(Z(rad(K(G/N))^*))$ resp. by a normal chain with Abelian factors.

2.5.9 Proposition

Let p be a prime number, G a finite p-group, N a normal subgroup of G, K a field and $char(K) = p$. The following statements are valid:

(i) $exp(Z(G)) \leq exp(Z(N)) \cdot exp(Z(G/N))$

(ii) For all $g \in Z(N) \setminus Z(G)$ the identity $o(\overline{g^G}) \leq exp(Z(N))$ is true.

(iii) For all $g \in G \setminus N$ such that $gN \in Z(G/N)$ and $g^{exp(Z(G/N))} \in Z(N)$ are true the identity $o(\overline{g^G}) \leq exp(Z(N)) \cdot exp(Z(G/N))$ is valid.

(iv) For all $g \in G \setminus N$ such that $gN \in Z(G/N)$ and $g^{exp(Z(G/N))} \in N \setminus Z(N)$ are valid the statement $o(\overline{g^G}) \leq exp(Z(G/N)) \cdot o(\overline{(g^{exp(Z(G/N))})^G})$ is true.

Proof. Let $n, f \in \mathbb{N}$ such that $p^n = exp(Z(N))$ and $p^f = exp(Z(G/N))$ are valid.

ad(i): $(Z(G)N)/N$ is a central subgroup of G/N, and thus $Z(G)^{(p^f)} \subseteq N \cap Z(G) \subseteq Z(N)$ is true. Therefor $(Z(G)^{(p^f)})^{(p^n)} = \{1_G\}$ is valid, and part (i) is proven.

ad(ii): Let $g \in Z(N) \setminus Z(G)$. Because of $g^{(p^n)} = 1_G$ we deduce $C_G(g) < G = C_G(g^{(p^n)})$, and by using theorem 2.4.8 we conclude part (ii).

ad(iii): Let $g \in G \setminus N$ such that $gN \in Z(G/N)$ and $g^{(p^f)} \in Z(N)$ are true. We deduce $g^{(p^f \cdot p^n)} = 1_G$, and theorem 2.4.8 is used to prove part (iii).

ad(iv): Let $g \in G \setminus N$ such that $gN \in Z(G/N)$ and $g^{(p^f)} \in N \setminus Z(N)$ are true. Let $r \in \mathbb{N}$ such that $p^r = o(\overline{(g^{(p^f)})^N})$ is valid. Based on theorem 2.4.8 we deduce $C_N(g^{(p^{f+r-1})}) < C_N(g^{(p^{f+r})})$, and thus $C_G(g^{(p^{f+r-1})}) < C_G(g^{(p^{f+r})})$ is valid. We use again theorem 2.4.8 to finish the proof.◇

2.5.10 Lemma

Let p be a prime number, G a finite p-group, N a normal subgroup of G, K a field, $char(K) = p$ and $g \in G \setminus N$ such that gN is not central in G/N. Let $s \in \mathbb{N}$ such that $p^s = o(\overline{(gN)^{G/N}})$ is true. An element $h \in G$ exists such that $[g^{(p^s)}, h] \in N$ and $[g^{(p^{s-1})}, h] \notin N$ are true. If we define $x_0 := [g^{(p^s)}, h]$ and $x_n := [x_{n-1}, g]$ for all $n \in \mathbb{N}$, then the following statements are valid:

(i) For all $n \in \mathbb{N}$ the statement $x_n \in N$ is true.

(ii) For $x_0 = 1_G$ the statement $o(\overline{g^G}) \leq p^s$ is valid.

(iii) For $x_0 \neq 1_G = x_1$ the statement $o(\overline{g^G}) \leq p^s \cdot o(x_0)$ is true.

(iv) For all $r \in \mathbb{N}$ such that $x_r \neq 1_G = x_{r+1}$ is valid the statement $o(\overline{g^G}) \leq o(x_r)$ is true.

Proof. We use theorem 2.4.8 to deduce $C_{G/N}(gN) = C_{G/N}(g^{(p^{s-1})}N) < C_{G/N}(g^{(p^s)}N)$. Hence, an element $h \in G$ exists such that $x_0 = [g^{(p^s)}, h] \in N$

End-commutable orderings and exponents

and $[g^{(p^{s-1})}, h] \notin N$ are valid.

ad(i): This statement is straightforward to verify.

ad(ii): Based on $x_0 = 1_G$ we conclude $h \in C_G(g^{(p^s)}) \setminus C_G(g^{(p^{s-1})})$, and thus statement (ii) is a consequence of theorem 2.4.8.

ad(iii): Let $x_0 \neq 1_G = x_1$.
Let $k \in \mathbb{N}$ such that $p^k = o(x_0)$ is true. Because of $x_1 = 1_G$ the element x_0 commutes per definition with g and thus also with $g^{(p^s)}$. We use theorem 1.3 of chapter III in [26] to deduce $1_G = x_0^{(p^k)} = [g^{(p^s)}, h]^{(p^k)} = [g^{(p^s \cdot p^k)}, h]$. We assume that h centralizes the element $g^{(p^{s+k-1})}$. Again by using theorem 1.3 of chapter III in [26] the identity $1_G = [g^{(p^{s+k-1})}, h] = [(g^{(p^s)})^{(p^{k-1})}, h] = [g^{(p^s)}, h]^{(p^{k-1})} = x_0^{(p^{k-1})}$ would be true which is contradicting $o(x_0) = p^k$. We conclude that part (iii) is true by using theorem 2.4.8.

ad(iv): Let $r \in \mathbb{N}$ such that $x_r \neq 1_G = x_{r+1}$ is valid.
Let $k \in \mathbb{N}$ such that $o(x_r^{-1}) = o(x_r) = p^k$ is true. Per definition the element x_r commutes with g, and thus x_r^{-1} commutes with g. Therefor we deduce – again by using theorem 1.3 in chapter III of [26] – the identity $1_G = (x_r^{-1})^{(p^k)} = [g, x_{r-1}]^{(p^k)} = [g^{(p^k)}, x_{r-1}]$. We assume that x_{r-1} centralizes the element $g^{(p^{k-1})}$. Another usage of theorem 1.3 in chapter III of [26] yields to the statement $1_G = [g^{(p^{k-1})}, x_{r-1}] = [g, x_{r-1}]^{(p^{k-1})} = x_r^{(p^{k-1})}$ contradicting $o(x_r^{-1}) = p^k$. Part (iv) is now deductable by using theorem 2.4.8. ◇

2.5.11 Remark

Let p a prime number, G a finite p-group, K a field and $char(K) = p$. For all $g \in G \setminus Z(G)$ such that $g^p \in G \setminus Z(G)$ is true we use theorem 2.4.8 to deduce the statement $o(\overline{g^G}) \leq p \cdot o(\overline{(g^p)^G})$. ◇

2.5.12 Lemma

Let p be a prime number, G a finite p-group, N a normal subgroup of G and K a field of $char(K) = p$. If N or G/N is Abelian or G/N of exponent p, then $exp(Z(rad(KG)^*)) \leq exp(Z(rad(KN)^*)) \cdot exp(Z(rad(K(G/N))^*))$ is valid.

Proof. If G/N is Abelian, then the proof is deductable by using proposition 2.5.9, proposition 2.4.7 and theorem 2.4.8.
If N is Abelian, the we use proposition 2.5.9, lemma 2.5.10, proposition 2.4.7 and theorem 2.4.8 to finish the proof.

If G^p is contained in N and $g \in G \setminus Z(G)$, then $g^p \in N$ is true.
If $g^p \notin Z(N)$, then we use proposition 2.5.4 and remark 2.5.11 to deduce the statement $o(\overline{g^G}) \leq p \cdot o(\overline{(g^p)^G}) \leq p \cdot o(\overline{(g^p)^N})$.
If $g^p \in Z(N)$, then $g^{(p \cdot exp(Z(N)))} = 1_G$ is true, and based on theorem 2.4.8 we deduce $o(\overline{g^G}) \leq p \cdot exp(Z(N))$. Propositions 2.5.9 and 2.4.7 as well as theorem 2.4.8 finish the proof.⋄

2.5.13 Theorem (bounding the exponent with chains)

Let p be a prime number, G a finite p-group, K a field, $char(K) = p$ and $\{1_G\} = N_r < N_{r-1} < \cdots < N_2 < N_1 = G$ a subnormal series of G such that for all $i \in \underline{r-1}$ the factor group N_i/N_{i+1} is Abelian. For all $i \in \underline{r-1}$ the statement $exp(Z(rad(KG)^*)) \leq exp(Z(rad(KN_i)^*)) \cdot \prod_{t=1}^{i-1} exp(N_t/N_{t+1})$ is true.

Proof. The proof is to be done by an induction argument based on r by using lemma 2.5.12 and part (iii) of proposition 2.5.2.⋄

2.5.14 Remark

Let p be a prime number, G a finite p-group, U a subgroup and N a normal subgroup of G. We ask whether $exp(Z(rad(KG)^*))$ can be bounded by $exp(Z(rad(KU)^*))$, $exp(Z(rad(KN)^*))$ or $exp(Z(rad(K(G/N))^*))$. This is not true and examples are presented at the end of chapter 3 (see Remarks to the bounds within section 2.5). In addition, we provide there some examples for lemma 2.5.12 and theorem 2.5.13.⋄

2.6 Open topics and exercises

Open-ended questions 2 *(i) How many end-commutable ordering does a finite nilpotent group and each conjugacy class possess? Can we bound the number of subnormal chains from 1 to G by this number?*

(ii) How can we construct all end-commutable orderings for a finite nilpotent group and for every conjugacy class of it?

(iii) How can we describe the action of the automorphism group of a finite nilpotent group of its end-commutable orderings resp. on all orderings of its conjugacy classes? What are the orbits of these actions?

(iv) Re-prove known theorems for nilpotent groups by using the concept of end-commutable orderings.

(v) Describe maximal end-commutable orderings of groups! Which ones are connected to Carter subgroups (see part (iii) of remark 2.3.11)?

End-commutable orderings and exponents

(vi) Is it possible to extend every end-commutable ordering of groups to a maximal end-commutable ordering to the left or right (see part (iii) of remark 2.3.11)?

(vii) Find the connection between the nilpotency class and the order of a conjugacy class sum.

Excercise 44 Prove corollary 2.3.8 by an induction argument.

Excercise 45 Enhance part (ii) of examples 2.1.2 to arbitrary prime numbers.

Excercise 46 Generalize part (v) of theorem 2.4.8 to an arbitrary element g of order $p^k \cdot n$ such that $k, n \in \mathbb{N}$ and $\gcd(n, p) = 1$ are true. Analyze on what terms $\overline{g^G}$ is nilpotent.

Excercise 47 Apply corollary 1.3.9 to the field $GF(2)$ and to the subsets $\{i\}, \{j\}, \{k\}$ and $\{i, j\}$ of Q_8. In addition, find a subset T of Q_8 such that $N_G(T) < N_G(\langle T \rangle_{\mathfrak{g}})$ is true.

Excercise 48 Let $T := \{(12), (34)\}$. Prove that $N_{S_4}(T)$ is no normal subgroup of $N_{S_4}(\langle T \rangle_{\mathfrak{g}})$.

Excercise 49 Let G be a finite nilpotent group. $\mid Aut(G) \mid$ is a divisor of $\mid EA(G) \mid$.

Excercise 50 Let K be a field of characteristic 2 and $G := Q_8$. Which subsets T of G are end-commutable? How many end-commutable ordering does each subset possess? Does a minimal $r \in \mathbb{N}$ exist such that $\overline{T}^{2^r} = 0$ is valid in KG for every subset T of G?

Excercise 51 Let K be a field of characteristic 2 and $G := D_8$. Which subsets T of G are end-commutable? How many end-commutable ordering does each subset possess? Does a minimal $r \in \mathbb{N}$ exist such that $\overline{T}^{2^r} = 0$ is valid in KG for every subset T of G?

Excercise 52 Let K be a field of characteristic 2 and $G := SD_8$. Which subsets T of G are end-commutable? How many end-commutable ordering does each subset possess? Does a minimal $r \in \mathbb{N}$ exist such that $\overline{T}^{2^r} = 0$ is valid in KG for every subset T of G?

Excercise 53 Let K be a field of characteristic 2 and $G := Q_{16}$. Determine an end-commutable ordering for G and for one non-central conjugacy X of G. Draw a picture for the determination process. Afterwards calculate the order of $\overline{G}$ and of $\overline{X}$ with respect to $*$ in KG. Compare each order with the corresponding nilpotency class of the element in $rad(KG)$.

Excercise 54 Let K be a field of characteristic 2 and $G := D_{16}$. Determine an end-commutable ordering for G and for one non-central conjugacy X of G. Draw a picture for the determination process. Afterwards calculate the order of $\overline{G}$ and of $\overline{X}$ with respect to $*$ in KG. Compare each order with the corresponding nilpotency class of the element in $rad(KG)$.

Excercise 55 Let K be a field of characteristic 2 and $G := SD_{16}$. Determine an end-commutable ordering for G and for one non-central conjugacy X of G. Draw a picture for the determination process. Afterwards calculate the order of $\overline{G}$ and of $\overline{X}$ with respect to $*$ in KG. Compare each order with the corresponding nilpotency class of the element in $rad(KG)$.

Excercise 56 Apply remark 2.3.10 to D_{32} and Q_8.

Excercise 57 Apply remark 2.3.10 to SD_{32} and SD_{32}.

Excercise 58 Apply remark 2.3.10 to Q_{32} and D_8.

Excercise 59 Calculate example 2.1.9 for D_8.

Excercise 60 Calculate example 2.1.9 for SD_8.

Excercise 61 Characterize nilpotent groups by end-commutable orderings.

Excercise 62 What is the definition of an end-commutable ordering?

Excercise 63 Prove that the group S_3 possesses subsets which are not end-commutable. Present two of these subsets explicitly.

Excercise 64 What is one of the most important property for end-commutable orderings and why?

Excercise 65 Prove remark 2.3.10 in details.

Excercise 66 Calculate the construction 2.2.4 for $G = Q_{16}$ and the conjugacy classes a^G, $(ab)^G$ and b^G. Present the results within a suitable graphic.

Excercise 67 Calculate the construction 2.2.4 for $G = D_{64}$ and the conjugacy classes a^G, $(ab)^G$ and b^G. Present the results within a suitable graphic.

Excercise 68 Let G be a finite p-group, U a subgroup of G and K a field of characteristic p. Determine the order of $\overline{U}$ with respect to $*$ in KG. (Tip: calculate the square of $\overline{U}$) Compare the order with the corresponding nilpotency class of the element in $rad(KG)$.

End-commutable orderings and exponents   65

Excercise 69 Let G be a finite p-group, $g \in G \setminus Z(G)$ and K a field of characteristic p. If g is of order p, then $\overline{g^G}$ is of order p with respect to $*$. What is the consequence for the involutions in G in the case $p = 2$?

Excercise 70 Let G be a finite nilpotent group and $g \in G \setminus Z(G)$. What is the consequence if G and the normal subgroup $N := \langle g^G \rangle_{\mathfrak{G}}$ are identical? What can be deduced if g^G is still a whole conjugacy class in N? (Tip: proposition 2.2.1)

Excercise 71 Let us focus on $G := Q_{32}$. For all conjugacy classes of G construct an end-commutable ordering and derive an end-commutable ordering for G based on the constructed orderings for these classes. Which subsets are normal in G? How can we determine end-commutable orderings for them? Visualize the results and the constructions!

Excercise 72 Let G be a finite p-group of exponent p and K a field of characteristic p. What is the order of the conjugacy class sums of G in $E(KG)$ and of the center $Z(G)$ of G? What is the consequence for the exponent and for the structure of $Z(1 + rad(KG))$? Construct an example for such a group G based on matrices and apply the results of this exercise to it!

Excercise 73 Determine φ_T for the following sets T:

(i) $T = \{a, b, c, d, \cdots, z\}$

(ii) $T = \{1, 3, 5, 7, 9, 11\}$

(iii) $T = \{2, 4, 6, 8, 10, 12\}$

(iv) $T = \{1, 2, 3, \cdots, 12\}$

(v) $T = \{2, 3, 5, 7, 11, 13, 17\}$.

Excercise 74 Let $T := \underline{4}$ and $i \in \underline{4}_0$. Determine $\binom{T}{i}$ explicitly.

Excercise 75 Prove proposition 2.4.3 in details.

Excercise 76 Apply proposition 2.4.3 to $n \in \underline{4}$.

Excercise 77 Apply part (i) of corollary 2.4.4 to $n \in \underline{4}$.

Excercise 78 Apply part (ii) of corollary 2.4.4 to $n = 3, p = 2$ and to $p = 3, n = 4$.

Excercise 79 Apply part (ii) of corollary 2.4.4 to the element $(a-1)+(b-1) \in rad(GF(2)D_8)$ and calculate all 2-powers of it with respect to $\cdot$ and $*$.

Excercise 80 *What is the content of the main theorem of the structure of finitely generated Abelian groups? Apply it to finite Abelian p-groups.*

Excercise 81 *Let G be a finite Abelian p-group and K a field of characteristic p. Prove that G and $1+rad(KG)$ possess the same exponent. (Tip: What are the p-powers of two commuting elements in the case $char(K) = p$?)*

Excercise 82 *Apply exercise 81 to an elementary Abelian group G and determine the structure of $1 + rad(KG)$. Decompose $1 + rad(KG)$ in cyclic groups.*

Excercise 83 *Prove remark 2.4.6 in details.*

Excercise 84 *What are the consequences of proposition 2.4.7 for $K = GF(8)$ and $G \in \{D_{16}, Q_{16}, SD_{16}\}$?*

Excercise 85 *What are the consequences of theorem 2.4.8 for $K = GF(8)$ and $G \in \{D_{16}, Q_{16}, SD_{16}\}$? For all these groups determine the exponent of the center of $1 + rad(KG)$.*

Excercise 86 *Apply the results of corollary 2.5.2, corollary 2.5.3, proposition 2.5.4 and corollary 2.5.5 to the groups D_{32}, Q_{32} and SD_{32} for the field $K := GF(64)$. Deduce results concerning the conjugacy class sums of the generating set $\{a, b\}$ of these groups. Use the subgroups generated by a resp. by b.*

Excercise 87 *Prove remark 2.5.11 in details.*

Excercise 88 *Apply theorem 2.5.12 to the derivation of G. What is the result for $G = SD_{64}$?*

Excercise 89 *Within SD_{32} construct adequate chains for subnormal subgroups and apply theorem 2.5.13 to them.*

Excercise 90 *Apply theorem 2.5.13 to the lower and upper central series and to the derived series.*

Excercise 91 *Let G be a finite p-group, K a field, $char(K) = p$ and $n \in \mathbb{N}$ such that p^n is the maximal length of all conjugacy classes of G. For every $i \in \underline{n}_0$ let C_{p^i} the union of all conjugacy classes of length p^i of G. Determine the order of $\overline{C_{p^i}}$ for all $i \in \underline{n}_0$. (Tip: G is the union of all conjugacy classes of G. Determine the p-th power of $\overline{G}$. Apply the theorem about the p-th power of conjugacy class sums.)*

Excercise 92 *Let G be a finite p-group, K a field, $char(K) = p$ and $n \in \mathbb{N}$ such that p^n is the maximal length of all conjugacy classes of G. For every $i \in \underline{n}_0$ let $C_{\leq p^i}$, $C_{\geq p^i}$, $C_{< p^i}$ and $C_{> p^i}$ the union of all conjugacy classes of length $\leq p^i$, $\geq p^i$, $< p^i$ and $> p^i$ of G. Apply exercise 91 to the corresponding sums in KG and determine their orders with respect to $*$.*

Chapter 3

The exponent of the center for special classes of groups

3.1 The maximal possible exponent

3.1.1 Proposition

Let p be a prime number, K a field, $char(K) = p$, G a finite p-group and M a maximal subgroup of G. For all $m \in Z(M) \setminus Z(G)$ the identity $o(\overline{m^G}) = o(mZ(G))$ is valid.

Proof. We use the assumptions to deduce that M is the centralizer of m in G. For all $r \in \mathbb{N}$ we conclude that $C_G(m) < C_G(m^{p^r})$ is valid if and only if m^{p^r} is central in G. Thus, the minimal $r \in \mathbb{N}$ such that $C_G(m) < C_G(m^{p^r})$ is true is exactly the order of $mZ(G)$ in $G/Z(G)$. We finish the proof by applying theorem 2.4.8.◇

3.1.2 Examples

Let K be a field and $char(K) = 2$.

(i) **Dihedral groups:**

Let $n \in \mathbb{N}_{\geq 3}$, $G := D_{2^n}$, $a, b \in G$, $G = \langle a, b \rangle_{\mathfrak{G}}$, $o(a) = 2^{n-1}$, $o(b) = 2$ and $a^b = a^{-1}$. $Z(G) = \langle a^{(2^{n-2})} \rangle_{\mathfrak{G}}$ is valid, and the subgroup $M := \langle a \rangle_{\mathfrak{G}}$ is Abelian and maximal in G. Because of $n \in \mathbb{N}_{\geq 3}$ we deduce $a \in Z(M) \setminus Z(G)$, and we use proposition 3.1.1 to calculate $o(\overline{a^G}) = o(aZ(G)) = 2^{n-2}$.

(ii) **Semi-dihedral groups:**

Let $n \in \mathbb{N}_{\geq 3}$, $G := SD_{2^n}$, $a, b \in G$, $G = \langle a, b \rangle_{\mathfrak{G}}$, $o(a) = 2^{n-1}$, $o(b) = 2$ and $a^b = a^{-1+2^{n-2}}$. An analogue argumentation as done within part (i) is used to prove $o(\overline{a^G}) = o(aZ(G)) = 2^{n-2}$.

(iii) **Quaternion groups:**
Let $n \in \mathbb{N}_{\geq 3}$, $G := Q_{2^n}$, $a, b \in G$, $G = \langle a, b \rangle_\mathfrak{g}$, $o(a) = 2^{n-1}$, $b^2 = a^{2^{n-2}}$ and $a^b = a^{-1}$. As done within part (i) we deduce $o(\overline{a^G}) = o(aZ(G)) = 2^{n-2}$.

(iv) **Enhancement of proposition 3.1.1:**
Let p be a prime number, G a finite p-group and $char(K) = p$. We use the propositions 3.1.1 and 2.5.4 to deduce an estimation for $o(\overline{g^G})$. Let $n \in \mathbb{N}$, for all $i \in \underline{n}$ let M_i be a maximal subgroup of M_{i+1} such that $M_{n+1} = G$ and $g \notin Z(M_i)$ are valid for all $i \in \underline{n+1} \setminus \underline{1}$ and $g \in Z(M_1)$. We deduce $o(\overline{g^G}) \leq o(\overline{g^{M_i}}) = o(gZ(M_i))$ (in $M_i/Z(M_i)$) for all $i \geq 2$.

Based on part (i) in the case $n = 4$ we apply this estimation to $\overline{b^G}$. Let $M_3 = G$, $M_2 = \langle a^2 \rangle_\mathfrak{g}$ and $M_1 = \langle a^4 \rangle_\mathfrak{g}$. $b \notin Z(M_3) \cup Z(M_2)$ and $b \in Z(M_1)$ are valid, and we deduce $o(\overline{b^G}) \leq o(\overline{b^{M_2}}) = o(bZ(M_2)) = 2$.◊

3.1.3 Remark

Let p be a prime number, K a field, $char(K) = p$ and G a non-Abelian finite p-group. If $exp(Z(G)) = \frac{|G|}{p^2}$ is valid, then we use proposition 2.4.7 and corollary 2.5.3 to deduce that $exp(Z(rad(KG)^*))$ is exactly the maximal possible upper bound $\frac{|G|}{p^2}$.◊

3.1.4 Proposition

Let p be a prime number, K a field, $char(K) = p$ and G a finite non-Abelian p-group. If G possesses a cyclic maximal subgroup, then the exponent of $Z(rad(KG)^*)$ is exactly $\frac{|G|}{p^2}$. In particular, $exp(Z(rad(KG)^*))$ is exactly the maximal possible upper bound $\frac{|G|}{p^2}$ (see corollary 2.5.3).

Proof. We use the classification on pages 98 and 99 in [71] and example 3.1.2. Therefor, we have only to analyze two types of groups which are presented in parts (a) and (d) of theorem 5.3.2 in [71].

<u>Case 1:</u> Let $p \neq 2$, $n \in \mathbb{N}_{\geq 3}$ and $h, a \in G$ such that $o(h) = p^n$, $o(a) = p$, $G = \langle h, a \rangle_\mathfrak{g}$, $|G| = p^{n+1}$ and $h^a = h^{1+p^{n-1}}$ are valid.
We use $(h^p)^a = (h^a)^p = (h^{1+p^{n-1}})^p = h^p$ and deduce $Z(G) = \langle h^p \rangle_\mathfrak{g}$. Thus, $exp(Z(G)) = \frac{|G|}{p^2}$ is true, and remark 3.1.3 finishes the proof for this case.

<u>Case 2:</u> We focus on the group G within part (d) of theorem 5.3.2 in [71]. This is the same group as analyzed in case 1 but in the case $p = 2$.◊

3.1.5 Lemma

Let p be a prime number, K a field, $char(K) = p$, $n \in \mathbb{N}_{\geq 4}$ and G a non-Abelian group such that $|G| = p^n$ is valid. For all $g \in G \setminus Z(G)$ such that $o(\overline{g^G}) = \frac{|G|}{p^2}$ is true the identity $o(g) = \frac{|G|}{p}$ is valid.

Proof. If $g \in G \setminus Z(G)$ such that $o(\overline{g^G}) = p^{n-2}$ is true, then we use theorem 2.4.8 to deduce

(1) $C_G(g) = C_G(g^{p^{n-3}}) < C_G(g^{p^{n-2}})$.

Thus, g is of order p^{n-1} or p^{n-2}, and we assume that

(2) $o(g) = p^{n-2}$

is valid. Based on (1) and (2) we would conclude

(3) $Z(G) \cap \langle g \rangle_\mathfrak{g} = \{1_G\}$.

Because of $\langle g \rangle_\mathfrak{g} Z(G) \leq C_G(g) < G$ the statements (1) and (2) would imply that $(Z(G), \langle g \rangle_\mathfrak{g})$ is a direct decomposition of $C_G(g)$, the center of G is of order p and the centralizer of g in G is of order p^{n-1}. In particular, $C_G(g)$ would be a normal subgroup of G, and the conjugacy classes of $g, \ldots, g^{p^{n-3}}$ would be all of length p. Based on $g \in C_G(g)$ we would deduce $g^G \subseteq C_G(g)$. Thus, $(g^{p^{n-3}})^G = (g^G)^{p^{n-3}} \subseteq C_G(g)^{p^{n-3}} = \langle g^{p^{n-3}} \rangle_\mathfrak{g}$ would be valid. The first and the last set of this statement are of order p. Therefor, 1_G and $g^{p^{n-3}}$ would be conjugated in G contradicting (2). ◇

Now we describe those finite p-groups G for which the exponent of $Z(rad(KG)^*)$ is of maximal value $\frac{|G|}{p^2}$.

3.1.6 Theorem (maximal possible exponent)

Let p be a prime number, K a field, $char(K) = p$ and G a non-Abelian finite p-group. The following statements are equivalent:

(i) $exp(Z(rad(KG)^*)) = \frac{|G|}{p^2}$

(ii) G possesses a cyclic maximal subgroup or $Z(G) \cong_\mathfrak{g} Z_{\frac{|G|}{p^2}}$ is valid.

Proof. The implication from (ii) to (i) is a consequence of proposition 3.1.4 and remark 3.1.3. If part (i) is true, then proposition 2.4.7 is used to deduce that $exp(Z(G)) = \frac{|G|}{p^2}$ is valid or an element $g \in G \setminus Z(G)$ exists such that $o(\overline{g^G}) = \frac{|G|}{p^2}$ is true. Part (ii) is now deductable by using remark 3.1.3 and lemma 3.1.5.◇

3.1.7 Remark

First we observe that both classes of groups in theorem 3.1.6 are not identical: Let p be an uneven prime number and G a non-Abelian finite p-group of order p^3 and of exponent p. $\mid Z(G) \mid = \frac{|G|}{p^2}$ is valid but G possesses no cyclic maximal subgroup. If $n \in \mathbb{N}_{\geq 4}$, then the dihedral group D_{2^n} possesses a cyclic maximal subgroup but the its center is of order $2 < \frac{|D_{2^n}|}{2^2}$.

Both classes of groups are well-known: p-groups possessing a cyclic maximal subgroup are classified on pages 98 and 99 in [71]. P-groups possessing a large center are classified in the forum mathoverflow in [77].◇

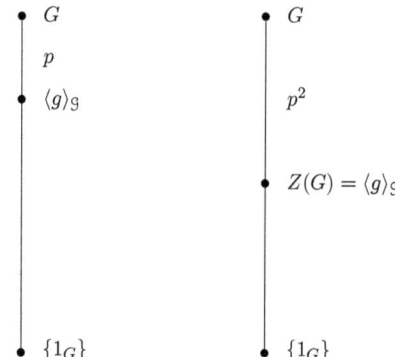

3.2 Elementary-Abelian centers

3.2.1 Theorem (elementary-Abelian centers)

Let p be a prime number, K a field, $char(K) = p$ and G a finite p-group. The following statements are equivalent:

(i) $Z(rad(KG)^*)$ is elementary Abelian.

(ii) $Z(G)$ is elementary Abelian and for all $g \in G \setminus Z(G)$ the condition $C_G(g) < C_G(g^p)$ is valid.

Proof. The proof is a consequence of proposition 2.4.7 and theorem 2.4.8.◇

The exponent of the center for special classes of groups

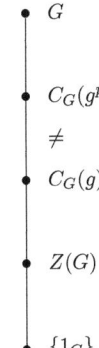

Every finite p-group G is $\mathcal{G}$-isomorphic to a subgroup of a p-Sylow subgroup of $GL(|G|, GF(p))$. In the following subsection we will prove that these groups satisfy the condition (ii) of theorem 3.2.1 (see also [5] where a similar result is proven).

3.2.2 The p-Sylow subgroups of $GL(n, GF(p^r))$

3.2.2.1 Remark

Let p be a prime number, $n \in \mathbb{N}$, K a finite field and $char(K) = p$. The group P_n of strict lower triangular matrices of $K^{n \times n}$ shifted by 1 is a p-Sylow subgroup of $GL(n, K)$.◇

3.2.2.2 Definition and remark (basis elements in matrix algebras)

Let K be a field and $n \in \mathbb{N}$. If $i, j \in \underline{n}$, then let $E_{i,j}$ be the $n \times n$-matrix such that the identities $(i;j)E_{i,j} = 1_K$ and $(k;l)E_{i,j} = 0_K$ are valid for all $k, l \in \underline{n}$ with $(k;l) \neq (i;j)$.
The set $\{E_{i,j} \mid i, j \in \underline{n}\}$ is a K-basis of $K^{n \times n}$, and for all $i, j, k, l \in \underline{n}$ with $j \neq k$ the identities $E_{i,j}E_{j,l} = E_{i,l}$ and $E_{i,j}E_{k,l} = 0_{K^{n \times n}}$ are true.
By $su(n, K)$ we denote the set of strict lower triangular matrices over K.◇

The following proposition is straightforward to verify:

3.2.2.3 Proposition

If p is a prime number, $n \in \mathbb{N}$, K a finite field and $char(K) = p$, then $Z(P_n) = 1_{K^{n \times n}} + \langle E_{n,1} \rangle_K$ is elementary-Abelian.◇

Based on the multiplication rule for matrices we can calculate:

3.2.2.4 Proposition

Let K be a field, $n \in \mathbb{N}$ and $A, B \in su(n, K)$. The following statements are valid:

(i) Let $r \in \underline{n}$ an element such that for all $i, j \in \underline{n} \setminus \underline{r-1}$ the identity $(i;j)A = (i;j)B = 0_K$ is true. Then for all $p \in \mathbb{N}$ and for all $i,j \in \underline{n} \setminus \underline{r-2}$ the condition $(i;j)(AB) = (i;j)(A^p) = 0_K$ is valid.

(ii) Let $r \in \underline{n}$ an element such that for all $i, j \in \underline{n} \setminus \underline{r-1}$ the condition $(i;j)A = 0_K$ is true. Then $E_{r,1}A = 0_K = AE_{r,1}$ is valid.

(iii) If $r \in \underline{n}$, then $AE_{r,1} = 0_K$ is true if and only if for all $i \in \underline{n}$ the condition $(i;r)A = 0_K$ is valid.⋄

3.2.2.5 Corollary (Sylow subgroups of linear groups)

Let p be a prime number, $n \in \mathbb{N}$, K a finite field and $char(K) = p$. $Z(rad(KP_n)^*)$ is elementary-Abelian.

Proof. We use theorem 2.4.8, proposition 2.4.7 and proposition 3.2.2.3 to deduce that we only have to prove that for all $g \in P_n \setminus Z(P_n)$ the group $C_{P_n}(g)$ is a proper subgroup of $C_{P_n}(g^p)$. Let $g \in P_n$. An element $M \in su(n, K)$ exists such that $g = 1_{K^{n \times n}} + M$ is valid. Because of $g^p = 1_{K^{n \times n}} + M^p$ we have to prove that $C_{su(n,K)}(M)$ is a proper subspace of $C_{su(n,K)}(M^p)$. Let us assume that both sets are identical.
We use $M \in su(n, K)$ and the conditions (i) and (ii) of proposition 3.2.2.4 to deduce that $M^p E_{n-1,1} = 0_{K^{n \times n}} = E_{n-1,1}M^p$ would be valid. By using the assumption and $E_{n-1,1}M = 0_{K^{n \times n}}$ we would conclude $ME_{n-1,1} = 0_{K^{n \times n}}$. Part (iii) of proposition 3.2.2.4 lets us prove that for all $i \in \underline{n}$ the identity $(i; n-1)M = 0_K$ would be true. Thus, part (i) of proposition 3.2.2.4 would be valid for $n-1$. We use $E_{n-2,1}$ and conclude by an analogue argumentation that $(i; n-2)M = 0_K$ would be true for all $i \in \underline{n}$. By using an induction argument we would finalize the argumentation with the result that $M = 0_{K^{n \times n}}$ is valid contradicting our assumption.⋄

3.2.2.6 Corollary (Sylow subgroups of GL, PGL, SL and PSL)

Let p be a prime number, $n \in \mathbb{N}$, K a finite field and $char(K) = p$. For every p-Sylow subgroup P of $GL(n, K)$, $PGL(n, K)$, $SL(n, K)$ and of $PSL(n, K)$ the center of $rad(KP)^*$ is elementary-Abelian.

Proof. Based on theorem 7.1 on page 185 in [26] the p-Sylow subgroups of these groups are all 𝒢-isomorphic. Thus, the proof is finished by using corollary 3.2.2.5. ◇

3.3 The minimal lower bound

For several interesting p-groups the exponent of the center of $1_G + rad(KG)$ is identical to the one of $Z(G)$. These are e.g. special, regular and minimal non-Abelian p-groups.

3.3.1 Proposition

If p is a prime number, K a field, $char(K) = p$ and G a finite p-group satisfying $exp(G/Z(G)) \leq exp(Z(G))$, then $exp(Z(rad(KG)^*)) = exp(Z(G))$ is valid.

Proof. The proof is a consequence of part (iii) of proposition 2.5.2.◇

3.3.2 Corollary

Let p be a prime number, K a field, $char(K) = p$ and G a finite p-group. The following statements are valid:

(i) If $cl(G) \leq 2$, then $exp(Z(rad(KG)^*)) = exp(Z(G))$ is valid.

(ii) If $|G'| \leq p$, then $exp(Z(rad(KG)^*)) = exp(Z(G))$ is true.

(iii) If G^p is central in G, then $exp(Z(rad(KG)^*)) = exp(Z(G))$ is valid.

(iv) If $\Phi(G)$ is central in G, then $exp(Z(rad(KG)^*)) = exp(Z(G))$ is true.

(v) If G is a special p-group, then $exp(Z(rad(KG)^*)) = p$ is valid.

Proof. ad(i): If $cl(G) \leq 2$ is valid, then we use part (a) of theorem 2.13 on page 266 in [26] to deduce $exp(G/Z(G)) \leq exp(Z(G))$. Thus, part (i) is a consequence of proposition 3.3.1.

ad(ii): This statement is a direct consequence of part (i).

ad(iii): If G^p is central in G, then $exp(G/Z(G)) \leq p \leq exp(Z(G))$ is valid. Hence, part (ii) is deductable based on proposition 3.3.1.

ad(iv): This statement is a direct consequence of part (iii).

ad(v): This statement is a direct consequence of part (iv).◇

3.3.3 Proposition (regular groups)

Let p be a prime number, G a finite regular p-group, K a field and $char(K) = p$. The following statements are valid:

(i) For all $g \in G \setminus Z(G)$ the identity $o(\overline{g^G}) = p$ is valid.

(ii) $exp(Z(rad(KG)^*)) = exp(Z(G))$

Proof. ad(i): Based on part (b) of theorem 10.6 on page 326 in [26] for all $a, b \in G$ the statement $[a^p, b] = 1_G$ is true if and only if $[a, b^p] = 1_G$ is valid. Part (i) is now a consequence of lemma 2.5.7 for $n = 1$.

ad(ii): This part is a direct consequence of part (i) and proposition 2.4.7.⋄

Now we prove that also all minimal non-Abelian finite p-groups satisfy the assumptions of part (iv) of corollary 3.3.2. In addition, we calculate the exponent of its center for all of its isomorphism types.

3.3.4 Proposition

If G is a minimal non-Abelian finite p-group, then $\Phi(G) = Z(G)$ is valid.

Proof. Let $g \in G \setminus Z(G)$ and U a maximal subgroup of G containing g. U is Abelian and centralizes g. Thus, $C_G(g) = U$ is valid because U is maximal and g is not central.
Let U be a maximal subgroup of G. If U would be central in G, then $Z(G)$ would be a maximal subgroup of G. G is finite and nilpotent, and thus $G/Z(G)$ is of order p. Let $x \in G$ such that $xZ(G)$ generates the factor group $G/Z(G)$. $G = Z(G) \cdot \langle x \rangle_{\mathfrak{g}}$ would be Abelian. Therefor, U is not central in G and an element $g \in U \setminus Z(G)$ exists. As already proven $U = C_G(g)$ is valid. We have proven that the maximal subgroups of G are exactly the centralizers of all non-central elements of G. The centralizer of all central elements in G is exactly G. We conclude
$$Z(G) = \bigcap_{g \in G} C_G(g) = \bigcap_{g \in G \setminus Z(G)} C_G(g) = \bigcap_{M < G, M \, maximal} M = \Phi(G).\diamond$$

3.3.5 Corollary (minimal non-Abelian groups)

If p is a prime number, G a minimal non-Abelian finite p-group, K a field and $char(K) = p$, then $exp(Z(rad(KG)^*)) = exp(Z(G))$ is true.

Proof. The corollary is a direct consequence of proposition 3.3.4 and part (iv) of corollary 3.3.2.⋄

3.3.6 Remark (isomorphism types of minimal non-Abelian groups)

Let p be a prime number and G a finite minimal non-Abelian p-group. Based on exercise 22 on page 309 in [26] the following three isomorphism types exist:

Type 1: $G \cong_g Q_8$:
In this case $exp(Z(Q_8)) = 2$ is valid (see examples 3.1.2).

Type 2: Elements $r \in \mathbb{N}_{\geq 2}$, $s \in \mathbb{N}$ and $a, b \in G$ exist such that $G = \langle a, b \rangle_g$, $o(a) = p^r$, $o(b) = p^s$ and $a^b = a^{1+p^{r-1}}$ are valid. $(\langle a \rangle_g, \langle b \rangle_g)$ is a semidirect decomposition of G. A straightforward calculation shows us that $(\langle a^p \rangle_g, \langle b^p \rangle_g)$ is a semidirect decomposition of $Z(G)$. Thus, $exp(Z(G)) = max\{p^{r-1}, p^{s-1}\}$ is true.

Type 3: Elements $r, s \in \mathbb{N}$ and $a, b \in G$ exist such that $G = \langle a, b \rangle_g$, $o(a) = p^s$, $o(b) = p^r$, $o([a, b]) = p$ and $\mid G \mid = p^{s+r+1}$ are valid. We prove that $G' = \langle [a, b] \rangle_g$ is true.
Let $g \in G \setminus Z(G)$. Every subgroup of G is Abelian, and thus $C_G(g)$ is of index p in G. Hence, every non-central conjugacy class of G is of length p. Based on a theorem of Knoche (see [36]) we deduce $\mid G' \mid = p$, and thus the first part is proven.
We proceed by deducing for $p \neq 2$ the identities $G^p = \langle a^p \rangle_g \langle b^p \rangle_g$ and $G^2 = G' \langle a^2 \rangle_g \langle b^2 \rangle_g$.
Let $z := [a, b] \in Z(G)$. $ab = baz$ is valid, and by using an induction argument we prove for all $j, k, n \in \mathbb{N}$ the statement

(1) $(a^j b^k)^n = z^{nk + \frac{n(n+1)j}{2}} b^{nk} a^{nj}$.

Let $g \in G$. Because of (1) and $G = \langle a, b \rangle_g$ elements $j \in \underline{p^s}$, $k \in \underline{p^r}$ and $i \in \underline{p}$ exist such that $g = z^i a^j b^k$ is true. We use (1) and $o(z) = p$ to deduce $g^p = (a^i b^j)^p = z^{\frac{p(p+1)j}{2}} b^{pk} a^{pj}$. For $p \neq 2$ we derive $g^p = b^{pk} a^{pj}$, and for $p = 2$ we prove $g^2 = z^{3j} b^{2k} a^{2j}$.
Proposition 3.3.4 is used to conclude $Z(G) = \langle a^p \rangle_g \langle b^p \rangle_g \langle z \rangle_g$, and thus the exponent of $Z(G)$ is exactly $max\{p^{s-1}, p^{r-1}, p\}$. ⋄

Within the next sections of this chapter we analyzed the exponent of $Z(rad(KG)^*)$ for diverse construction of groups (central products, direct products, wreath products and group extensions).

3.4 Central products

We begin this section by defining central products and by presenting some basic facts for them.

3.4.1 Definition (central products)

Let G_1, G_2 be groups, $U_i \leq Z(G_i)$ for all $i \in \underline{2}$ and $\mu : U_1 \longrightarrow U_2$ a $\mathcal{G}$-isomorphism. $D_\mu := \{(u; (u\mu)^{-1}) \mid u \in U_1\}$ is a central subgroup of $G_1 \times G_2$. We define $G_1 \mathsf{Y}_\mu G_2 := (G_1 \times G_2)/D_\mu$ and call this group the central product of G_1 and G_2 based on the unified central subgroups U_1 and U_2 with respect to μ. If μ is well-known, then we use also the symbols D and $G_1 \mathsf{Y} G_2$ instead of D_μ and $G_1 \mathsf{Y}_\mu G_2$. ⋄

The next theorem justifies this definition:

3.4.2 Theorem

Let G_1, G_2 be groups, $U_i \leq Z(G_i)$ for all $i \in \underline{2}$ and $\mu : U_1 \longrightarrow U_2$ a $\mathcal{G}$-isomorphism. The following statements are valid:

(i) $\alpha_1 : G_1 \longrightarrow G_1 \mathsf{Y} G_2$, $g \mapsto (g; 1_{G_2})D$ is a $\mathcal{G}$-monomorphism.

(ii) $\alpha_2 : G_2 \longrightarrow G_1 \mathsf{Y} G_2$, $g \mapsto (1_{G_1}; g)D$ is a $\mathcal{G}$-monomorphism.

(iii) $G_1\alpha_1$ and $G_2\alpha_2$ are two normal subgroups of G such that $[G_1\alpha_1, G_2\alpha_2] = 1$ and $G_1 \mathsf{Y} G_2 = (G_1\alpha_1) \cdot (G_2\alpha_2)$ are valid.

(iv) For all $u \in U_1$ the condition $u\alpha_1 = u(\mu\,\alpha_2)$ is valid.

(v) $(G_1\alpha_1) \cap (G_2\alpha_2) = U_1\alpha_1 = U_2\alpha_2$.

Proof. see theorem 3.10 on page 49 in [26]. ⋄

Central products satisfy the following symmetry property:

3.4.3 Remark

Let G_1, G_2 be groups, $U_i \leq Z(G_i)$ for all $i \in \underline{2}$ and $\mu : U_1 \longrightarrow U_2$ a $\mathcal{G}$-isomorphism. The function

$$\Phi : G_1 \mathsf{Y}_\mu G_2 \longrightarrow G_2 \mathsf{Y}_{\mu^{-1}} G_1, \ (g_1; g_2)D_\mu \mapsto (g_2; g_1)D_{\mu^{-1}}$$

is a $\mathcal{G}$-isomorphism. ⋄

The exponent of the center for special classes of groups   77

3.4.4   Proposition

Let G_1, G_2 be groups, $U_i \leq Z(G_i)$ for all $i \in \underline{2}$ and $\mu : U_1 \longrightarrow U_2$ a $\mathcal{G}$-isomorphism. The following statements are valid:

(i) $Z(G_1 Y G_2) = (Z(G_1) \times Z(G_2))/D = Z(G_1) Y Z(G_2)$

(ii) If p is a prime number and G_1, G_2 finite p-groups, then
$exp(Z(G_1 Y G_2)) = max\{exp(Z(G_1)), exp(Z(G_2))\}$ is valid.

Proof. ad(i): It is straightforward to prove that $(Z(G_1) \times Z(G_2))/D$ is central in $G_1 Y G_2$. Let $(g_1; g_2) \in G_1 \times G_2$ such that $(g_1; g_2)D$ is central in $G_1 Y G_2$. If $a \in G_1$, then $(g_1; g_2)D(a; 1_{G_2})D = (a; 1_{G_2})D(g_1; g_2)D$ and $([g_1, a]; 1_{G_2})D = D$ are valid. Based on (i) of theorem 3.4.2 we deduce $[g_1, a] = 1_{G_1}$, and g_1 is central in G_1. If we apply this result to remark 3.4.3, then we conclude that g_2 is central in G_2. The second identity in (ii) is valid per definition 3.4.1.

ad(ii): Based on (i) and parts (i) and (ii) of theorem 3.4.2 we derive that $exp(Z(G_1 Y G_2)) \geq max\{exp(Z(G_1)), exp(Z(G_2))\}$ is true. $Z(G_1 Y G_2)$ is an Abelian group, and thus (i) and part (iii) of theorem 3.4.2 are finishing the proof.⋄

3.4.5   Proposition

Let G_1, G_2 be groups, $U_i \leq Z(G_i)$ for all $i \in \underline{2}$ and $\mu : U_1 \longrightarrow U_2$ a $\mathcal{G}$-isomorphism. The following statements are valid:

(i) If g_2 is central in G_2, then $C_{G_1 Y G_2}((g_1; g_2)D) = (C_{G_1}(g_1)\alpha_1)(G_2 \alpha_2)$ is valid.

(ii) If g_1 is central in G_1, then $C_{G_1 Y G_2}((g_1; g_2)D) = (G_1 \alpha_1)(C_{G_2}(g_2)\alpha_2)$ is valid.

Proof. Because of remark 3.4.3 we only need to prove part (i). $(C_{G_1}(g_1)\alpha_1)(G_2 \alpha_2)$ centralizes the element $(g_1; g_2)D$. Let $(a; b) \in G_1 \times G_2$ such that $(a; b)D \in C_{G_1 Y G_2}((g_1; g_2)D)$ is true. An element $u \in U_1$ exists such that $[a, g_1] = u$ and $[b, g_2] = (u\mu)^{-1}$ are valid. The element g_2 is central in G_2, and thus $(u\mu)^{-1} = 1_{G_2}$ is true. Therefor $u = 1_{G_1}$ and $a \in C_{G_1}(g_1)$ are valid.⋄

3.4.6   Lemma

Let p be a prime number, K a field, $char(K) = p$, G_1, G_2 finite p-groups, $U_i \leq Z(G_i)$ for all $i \in \underline{2}$ and $\mu : U_1 \longrightarrow U_2$ a $\mathcal{G}$-isomorphism. The following statements are valid:

(i) If $g_2 \in Z(G_2)$ and $g_1 \notin Z(G_1)$ are valid, then
$$o(\overline{((g_1; g_2)D)}^{G_1 Y G_2}) = o(\overline{g_1}^{G_1})$$ is true.

(ii) If $g_1 \in Z(G_1)$ and $g_2 \notin Z(G_2)$ are valid, then

$$o(\overline{((g_1;g_2)D)^{G_1 Y G_2}}) = o(\overline{g_2^{G_2}}) \text{ is true.}$$

(iii) If $g_1 \notin Z(G_2)$ and $g_2 \notin Z(G_2)$ are valid, then

$$o(\overline{((g_1;g_2)D)^{G_1 Y G_2}}) \leq min\{o(\overline{g_1^{G_1}}), o(\overline{g_2^{G_2}})\} \text{ is true.}$$

Proof. ad(i): For all $n \in \mathbb{N}$ we calculate $((g_1;g_2)D)^n = (g_1^n;g_2^n)D$. Based on proposition 3.4.5 we deduce for all $n \in \mathbb{N}$ the identity $C_{G_1 Y G_2}(((g_1;g_2)D)^n) = (C_{G_1}(g_1^n)\alpha_1)(G_2\alpha_2)$. Because of theorem 2.4.8 we have only to prove that for all $n \in \mathbb{N}$ the condition (1) $C_{G_1}(g_1^n) > C_{G_1}(g_1)$ is valid if and only if the identity (2) $(C_{G_1}(g_1^n)\alpha_1)(G_2\alpha_2) > (C_{G_1}(g_1)\alpha_1)(G_2\alpha_2)$ is true. The implication from (2) to (1) is straightforward to be proven based on a contraposition argument. Let (1) be valid, and let $x \in C_{G_1}(g_1^n) \setminus C_{G_1}(g_1)$. $(x;1_{G_2})D$ centralizes the element $(g_1^n;g_2)D$. We assume that $(x;1_{G_2})D$ would centralize $(g_1;g_2)D$. Then an element $u \in U_1$ would exists such that $[x,g_1] = u$ and $1_{G_2} = [1_{G_2},g_2] = (u\mu)^{-1}$ would be valid. We would deduce $u = 1_{G_1}$ and $[x,g_1] = 1_{G_1}$ contradicting our assumption.

ad(ii): This statement is a consequence of part (i) and of remark 3.4.3.

ad(iii): We prove $o(\overline{((g_1;g_2)D)^{G_1 Y G_2}}) \leq o(\overline{g_1^{G_1}})$, and by using remark 3.4.3 the proof is finished. Let $n \in \mathbb{N}$ such that $o(\overline{g_1^{G_2}}) = p^n$ is valid. We use theorem 2.4.8 to deduce the existence of an element $x \in C_{G_1}(g_1^{p^n}) \setminus C_{G_1}(g_1)$. $(x;1_{G_2})D$ centralizes the element $((g_1;g_2)D)^{p^n}$, and we assume that $(x;1_{G_2})D$ centralizes $(g_1;g_2)D$. We would conclude $([g_1,x];1_{G_2})D = D$, and based on part (i) of theorem 3.4.2 we would conclude $[g_1,x] = 1_{G_1}$. This statement is contradicting the choice of x, and by using theorem 2.4.8 the lemma is proven.⋄

3.4.7 Theorem (central products)

Let p be a prime number, K a field, $char(K) = p$, G_1, G_2 finite p-groups, $U_i \leq Z(G_i)$ for all $i \in \underline{2}$ and $\mu : U_1 \longrightarrow U_2$ a $\mathcal{G}$-isomorphism. The exponent of $Z(rad(K(G_1 Y G_2))^*)$ is exactly

$$max\{exp(Z(rad(KG_1)^*)), exp(Z(rad(KG_2)^*))\}.$$

Proof. The theorem is proven by using proposition 2.4.7, theorem 2.4.8, proposition 3.4.4 and lemma 3.4.6.⋄

3.4.8 Corollary (direct products)

Let p be a prime number, K a field, $char(K) = p$ and G_1, G_2 finite p-groups. The exponent of $Z(rad(K(G_1 \times G_2))^*)$ is exactly

$$max\{exp(Z(rad(KG_1)^*)), exp(Z(rad(KG_2)^*))\}.$$

Proof. The corollary is a special case of theorem 3.4.7.◊

3.4.9   Example (Hamiltonian p-groups)

The non-Abelian Hamiltonian p-groups are – based on theorem 7.12 on page 308 in [26] – exactly the 2-groups G for which an element $n \in \mathbb{N}$ exists such that $G \cong_\mathcal{G} Q_8 \times Z_2^n$ is valid. If K is a field and $char(K) = 2$, then we use corollary 3.4.8 and the examples 3.1.2 to deduce that $Z(rad(KG)^*)$ is elementary-Abelian.◊

3.4.10   Corollary (extra-special p-groups)

Let p be a prime number, K a field, $char(K) = p$ and G a extra-special finite p-group. $Z(rad(KG)^*)$ is elementary-Abelian.

Proof. Every extra-special finite p-group is $\mathcal{G}$-isomorph to a central product of several non-Abelian groups of order p^3 (see e.g. [26]). Thus, the corollary is proven based on theorem 3.4.7 and corollary 2.5.3.◊

3.4.11   Further applications

(i) Let p be a prime number, K a field, $char(K) = p$ and G a non-Abelian p-group such that **every Abelian characteristic subgroup is cyclic**. Based on theorem 11.10 on page 357 in [26] the structure of G is limited to the following five cases:

<u>Case 1:</u> Let $p \neq 2$ be valid. G is a central product of a extra-special p-group and the cyclic group $Z(G)$. We use theorem 3.4.7 and corollary 3.4.10 to deduce that the exponent of $Z(rad(KG)^*)$ is exactly $\mid Z(G) \mid$.

<u>Case 2:</u> $p = 2$ is valid and G is an extra-special 2-group. Based on corollary 3.4.10 we derive that $Z(rad(KG)^*)$ is elementary-Abelian.

<u>Case 3:</u> G is a dihedral, semi-dihedral or a quaternion group. We use the examples 3.1.2 to deduce $exp(Z(rad(KG)^*)) = \frac{|G|}{2^2}$.

<u>Case 4:</u> $p = 2$ is valid and G is a central product of a extra-special 2-group and a cyclic 2-group Q. Corollary 3.4.10 and theorem 3.4.7 let us deduce that $exp(Z(rad(KG)^*))$ is exactly $\mid Q \mid$.

<u>Case 5:</u> Let $p = 2$ and G the central product of a extra-special 2-group and a dihedral, semi-dihedral or quaternion group of order 2^n. Based on

theorem 3.4.7, examples 3.1.2 and corollary 3.4.10 we deduce that the exponent of $Z(rad(KG)^*)$ is exactly 2^{n-2}.

(ii) Let p be a prime number, $p > 3$, K a field, $char(K) = p$ and G a non-Abelian p-group of order p^n such that **every Abelian normal subgroup can be $\mathcal{G}$-generated by 1 or 2 elements**. Theorem 12.4 on page 343 in [26] classifies these groups:

Case 1: G is **metacyclic**. Because of $p \neq 2$ the group G is – based on theorem 10.2 on page 322 in [26] – a regular p-group. We use proposition 3.3.3 to deduce $exp(Z(rad(KG)^*)) = exp(Z(G))$.

Case 2: G is a central product of a non-Abelian p-group of order p^3 and a cyclic group of order p^{n-2}. Based on theorem 3.4.7 and corollary 3.4.10 we derive $exp(Z(rad(KG)^*)) = p^{n-2}$.

Case 3: Elements $x, y, z \in G$ exist such that $G = \langle x, y, z \rangle_\mathcal{G}$, $o(x) = o(y) = p$, $o(z) = p^{n-2}$, $y^x = yz^{sp^{n-3}}$ and $z^x = yz$ are valid. In addition, $n \geq 4$ and s a quadratic nonresidue modulo p: p does not divide s and for no element $a \in \mathbb{N}$ the congruence $a^2 \equiv s \bmod p$ is valid. $(\langle y, z \rangle_\mathcal{G}, \langle x \rangle_\mathcal{G})$ is a semidirect decomposition of G and the normal subgroup $\langle y, z \rangle_\mathcal{G} = C_G(G')$ is Abelian. We determine the center of G. $Z(G) \subseteq C_G(G') = \langle y \rangle_\mathcal{G} \langle z \rangle_\mathcal{G}$ is straightforward to verify. Let $i \in \underline{p}$ and $j \in \underline{p^{n-2}}$. The element $g := y^i z^j$ is central if g is commuting with x. We calculate:
$g^x = g \iff (y^i z^j)^x = y^i z^j \iff (yz^{sp^{n-3}})^i (yz)^j = y^i z^j \iff z^{isp^{n-3}} y^j = 1$
$\iff p \mid j \wedge p^{n-2} \mid isp^{n-3} \iff p \mid j \wedge p \mid i$.
Thus, $Z(G) = \langle z^p \rangle_\mathcal{G}$ is valid.
We prove that for all $g \in G \setminus Z(G)$ the statement $C_G(g) < C_G(g^p)$ is true. Let $g \in G \setminus Z(G)$. $g^p \in C_G(G')$ is valid, and hence g^p is centralized by z. We assume that z is centralized by g, too. En element $k \in \underline{p-1}$ would exists such that $g \in C_G(G')x^i$ would be valid. Thus, z would centralize x^i and therefor also x which is a contradiction to our assumption. We conclude by using proposition 2.4.7 and theorem 2.4.8 that $exp(Z(rad(KG)^*)) = p^{n-3}$ is valid.⋄

3.5 Wreath products

We begin with a preliminary remark which is used for the definition of wreath products. The proof maybe executed as an exercise.

The exponent of the center for special classes of groups 81

3.5.1 Proposition

Let X be a set, H and S groups. S is acting on X by the action δ. For all $\varphi \in H^X$ and $s \in S$ let $\varphi^s : X \longrightarrow H$, $x \mapsto (x(s^{-1}\delta))\varphi$. The following statements are valid:

(i) For all $s \in S$ the function $\tilde{s} : H^X \longrightarrow H^X$, $\varphi \mapsto \varphi^s$ is an automorphism of H^X.

(ii) $f : S \longrightarrow Aut(H^X)$, $s \mapsto \tilde{s}$ is a $\mathcal{G}$-homomorphism. ⋄

3.5.2 Definition (wreath products)

Let H and S are groups. S is acting on a set X by the action δ. We use the $\mathcal{G}$-homomorphism f of part (ii) within proposition 3.5.1 and construct the semi-direct product of H^X and S based on f which we call the wreath product of S and H based on the action δ. We symbolize the wreath product by $H \wr_\delta S$ resp. $H \wr_X S$. We assume that S and for every subset T of X the set H^T is contained in $H \wr_X S$.

If $X = S$ and δ the right multiplication of S on S, then we call this wreath product the regular wreath product of S and H and symbolize this group by $H \wr S$. ⋄

3.5.3 Remark

Let H, S groups and X a set acting on S by δ. If H is the trivial group, then the centers of $H \wr_X S$ and $Z(S)$ are $\mathcal{G}$-isomorphic. ⋄

3.5.4 Proposition

Let H, S be groups, H non-trivial, X a finite set acting on S by δ and $B_1, \ldots, B_r$ the orbits of X under S. If $N := \{\varphi \mid \varphi \in H^X, \forall i \in \underline{r} \exists h_i \in Z(H) : \varphi_{|B_i} \equiv h_i\}$, then the following statements are valid:

(i) N is a central subgroup of $H \wr_X S$.

(ii) $Z(H \wr_X S) = N \cdot (Z(S) \cap ker\delta)$.

(iii) $Z(H \wr_X S) \cong_\mathcal{G} Z(H)^{\underline{r}} \times (Z(S) \cap ker\delta)$.

Proof. We have only to prove that the center of $H \wr_X S$ is contained in $N \cdot (Z(S) \cap ker\delta)$. Let $z \in Z(H \wr_X S)$. (H^X, S) is a semi-direct decomposition of $H \wr_X S$, and thus elements $\varphi \in Z(H^X)$ and $s \in Z(S)$ exist such that $z = \varphi s$ is valid.

Let us assume that an element $x \in X$ would exists such that $xs^{-1} \neq x$ would be valid. For an element $h \in H \setminus \{1_H\}$ we define $\alpha : X \longrightarrow H$ by $x \mapsto 1_H$ and $y \mapsto h$ for all $y \in X \setminus \{x\}$. φ is central in H^X, and thus we

conclude $\alpha = (\alpha)^{\varphi s} = \alpha^s$ and $1_H = x\alpha = x\alpha^s = (xs^{-1})\alpha = h$. This is a contradiction to our assumptions. Therefor, $s \in Z(S) \cap ker\delta$ is valid.
Let $i \in \underline{r}$ and $x_1, x_2 \in B_i$. An element $t \in S$ exists such that $x_2 = x_1 t$ is true. We use $\varphi = \varphi^{t^{-1}}$ to derive $x_1 \varphi = x_1 \varphi^{t^{-1}} = (x_1 t)\varphi = x_2 \varphi$, and thus the proposition is proven.◇

3.5.5 Corollary

Let p be a prime number, H, S p-groups and X a finite set on which S acts by δ. The following statements are valid:

(i) $exp(Z(H \wr_X S)) = max\{exp(Z(H)), exp(Z(S) \cap ker\delta)\}$

(ii) $exp(Z(H)) \leq exp(Z(H \wr_X S)) \leq max\{exp(Z(H)), exp(Z(S))\}$

Proof. Statement (i) is a consequence of part (ii) of proposition 3.5.4 and part (ii) is deductable by (i).◇

The main topic within our analysis for wreath products is the determination of the orders of conjugacy class sums for elements outside the normal subgroup H^X. This will be done within the next lemmas, and we will prove that these orders are small and not relevant for the calculation of the exponent.

3.5.6 Lemma

Let p be a prime number, H, S p-groups and S is acting transitive by δ on a set X. Let $s \in S$ such that $S = \langle s \rangle_{\mathcal{G}}$ and $s \notin ker\delta$ are valid. Then an element $\alpha \in Z(H^X)$ exists such that $\alpha^s \neq \alpha = \alpha^{s^p}$ is true.

Proof. S is acting transitive on X, and thus we can assume that a subgroup U of S exists such that $X = S/_r U$ is valid and δ is the right multiplication from S on the right cosets of U in S. Because of $s \notin ker\delta$ we deduce $U \subseteq \langle s^p \rangle_{\mathcal{G}}$.
Let R be system of representatives for the right cosets of U in S containing $\{1_S, s\} \subseteq R$ and $1_H \neq h$ an element of the center of H. We define $\alpha : S/_r U \longrightarrow H$ by $Ur \mapsto 1_H$ resp. $Ur \mapsto h$ for all $r \in R$ such that $r \in \langle s^p \rangle_{\mathcal{G}}$ resp. $r \notin \langle s^p \rangle_{\mathcal{G}}$ is valid, and by definition $\alpha \in Z(H^X)$ is true. We use $s \notin \langle s^p \rangle_{\mathcal{G}}$ to conclude $(Us)\alpha = h$, and because of $(Us)\alpha^s = (U1_S)\alpha = 1_H$ the statement $\alpha \neq \alpha^s$ is deduced. Let $r \in R \cap \langle s^p \rangle_{\mathcal{G}}$. By definition $(Ur)\alpha = 1_H$ is valid. If $a \in R$ and $Urs^{-p} = Ua$ are true, then $a \in \langle s^p \rangle_{\mathcal{G}}$ and $(Ur)\alpha^{s^p} = 1_H$ are valid. Let $r \in R \setminus \langle s^p \rangle_{\mathcal{G}}$. By definition of α we conclude $(Ur)\alpha = h$. If $a \in R$ and $Urs^{-p} = Ua$ are true, then $a \notin \langle s^p \rangle_{\mathcal{G}}$ is valid, and thus $(Ur)\alpha^{s^p} = h$ is true.◇

The exponent of the center for special classes of groups

3.5.7 Lemma

Let p be a prime number, K a field, $char(K) = p$, H, S p-groups and $S = \langle s \rangle_{\mathfrak{g}}$ is acting non-trivial by δ on a finite set X. For all $\varphi \in H^X$ the identity $o(\overline{(\varphi s)^{H \wr_X S}}) = p$ is valid.

Proof. Let $B_1, \ldots, B_r$ be the S-orbits of X. It is straightforward to prove that $Fix_X(s^{-1})$ is invariant under $\langle s \rangle_{\mathfrak{g}}$. W.l.o.g. we assume that an element $t \in \underline{r}$ exists such that $Fix_X(s^{-1})$ is the disjoint union of the orbits $B_1, \ldots, B_t$. s is not contained in the kernel of the action, and thus $t < r$ is valid. We deduce for the wreath product $H^{B_r} S$ that S is acting transitive on B_r. Based on lemma 3.5.6 an element $\alpha \in Z(H^{B_r})$ exists such that s^p but not s is centralized by it. Let $\varphi \in H^X$. We prove that $(\varphi s)^p$ but not φs is centralized by α. For all $i \in \underline{r}$ let $\varphi_i \in H^{B_i}$ such that $\varphi = \varphi_1 \ldots \varphi_r$ is true. Let us assume that φs is centralized by α. The argumentation

$(\varphi s)^\alpha = \varphi s \Longrightarrow$
$(\varphi_1 \ldots \varphi_{r-1} \varphi_r s)^\alpha = \varphi s \Longrightarrow$ $\quad\quad ([H^{B_i}, H^{B_j}] = \{1_{H^X}\}$ for all $i \neq j)$
$\varphi_1 \ldots \varphi_{r-1} \varphi_r^\alpha s^\alpha = \varphi s \Longrightarrow$ $\quad\quad\quad\quad\quad\quad\quad\quad\quad\quad\quad$ (choice of α)
$s^\alpha = s$

is contradicting the choice of α. We proceed by an induction argument that an element $\psi \in H^X$ exists such that $(\varphi s)^p = \psi s^p$ is true. For all $i \in \underline{r}$ let $\psi_i \in H^{B_i}$ such that $\psi = \psi_1 \ldots \psi_r$ is valid. We calculate

$(\psi s^p)^\alpha$
$= (\psi_1 \ldots \psi_{r-1} \psi_r s^p)^\alpha$
$= \psi_1 \ldots \psi_{r-1} \psi_r^\alpha (s^p)^\alpha$
$= \psi (s^p)^\alpha$
$= \psi s^p.$

Thus, $C_{H \wr_X S}(\varphi s) < C_{H \wr_X S}((\varphi s)^p)$ is true, and based on theorem 2.4.8 the proof is finished.◇

3.5.8 Corollary

Let p be a prime number, K a field, $char(K) = p$, H, S p-groups and S is acting by δ on a finite set X. For all $s \in S \setminus ker\delta$ and $\varphi \in H^X$ the identity $o(\overline{(\varphi s)^{H \wr_X S}}) = p$ is valid.

Proof. Let $s \in S \setminus ker\delta$ and $\varphi \in H^X$. The wreath product $H^X \langle s \rangle_{\mathfrak{g}}$ can be analyzed by lemma 3.5.7, and we use proposition 2.5.4 to deduce $o(\overline{(\varphi s)^{H \wr_X S}}) \leq o(\overline{(\varphi s)^{H \wr_X \langle s \rangle_{\mathfrak{g}}}}) = p$.◇

3.5.9 Proposition

Let p be a prime number, K a field, $char(K) = p$, H, S p-groups and S is acting by δ on a finite set X. For all $h \in H$ let $\alpha_h : X \longrightarrow H$, $x \mapsto h$. The following statements are valid:

(i) For all $h \in H \setminus Z(H)$ the identity $o(\overline{(\alpha_h)^{H \wr_X S}}) = o(\overline{h^H})$ is valid.

(ii) For all $s \in ker\delta \setminus Z(S)$ the identity $o(\overline{s^{H \wr_X S}}) = o(\overline{s^S})$ is true.

Proof. ad(i): Let $h \in H \setminus Z(H)$. For all $n \in \mathbb{N}$ the identities $(\alpha_h)^n = \alpha_{h^n}$ and $C_{H \wr_X S}((\alpha_h)^n) = C_H(h^n)^X S$ are valid. We use theorem 2.4.8 to prove statement (i).

ad(ii): Let $s \in ker\delta \setminus Z(S)$. For all $n \in \mathbb{N}$ the statement $C_{H \wr_X S}(s^n) = H^X C_S(s^n)$ is true, and by using theorem 2.4.8 we finalize the proof. $\diamond$

3.5.10 Lemma

Let p be a prime number, K a field, $char(K) = p$, H, S p-groups and S is acting by δ on a finite set X. Let $\varphi \in H^X$ and $s \in ker\delta$ such that φs is not central in $H \wr_X S$. The following statements are valid:

(i) If φ is central in $H \wr_X S$, then $s \in ker\delta \setminus Z(S)$ and $o(\overline{(\varphi s)^{H \wr_X S}}) \leq o(\overline{s^S})$ are valid.

(ii) If $\varphi \notin Z(H^X)$ is valid, then $o(\overline{(\varphi s)^{H \wr_X S}}) \leq o(\overline{\varphi^{H^X}})$ and $o(\overline{(\varphi s)^{H \wr_X S}}) \leq exp(Z(rad(KH)^*))$ are true.

(iii) If $\varphi \in Z(H^X) \setminus Z(H \wr_X S)$ and $s \in Z(S)$ are valid, then $o(\overline{(\varphi s)^{H \wr_X S}}) \leq o(\varphi)$ is true.
In particular, $o(\overline{(\varphi s)^{H \wr_X S}}) \leq exp(Z(H))$ is true.

(iv) If $\varphi \in Z(H^X) \setminus Z(H \wr_X S)$ and $s \notin Z(S)$ are valid, then $o(\overline{(\varphi s)^{H \wr_X S}}) \leq max\{o(\varphi), o(\overline{s^S})\}$ is true.
In particular, $o(\overline{(\varphi s)^{H \wr_X S}}) \leq max\{exp(Z(H)), o(\overline{s^S})\}$ is valid.

Proof. ad(i): If s would be central in S, then we would deduce by using $s \in ker\delta$ that s and φs are central in $H \wr_X S$. Thus, $s \in ker\delta \setminus Z(S)$ is true. Let $r \in \mathbb{N}$ such that $o(\overline{s^S}) = p^r$ is valid. Based on theorem 2.4.8 an element a in $C_S(s^{p^r}) \setminus C_S(s)$ exists. φ is central in $H \wr_X S$, and thus $(\varphi s)^a = \varphi s^a$ is true. a is not centralized by φs. In addition,

$$((\varphi s)^{p^r})^a \qquad \qquad (\varphi \text{ is central})$$
$$= ((\varphi)^{p^r}(s^{p^r}))^a \qquad (\varphi \text{ is central})$$
$$= (\varphi)^{p^r}(s^{p^r})^a \qquad \text{(choice of } a\text{)}$$

$$= (\varphi)^{p^r} s^{p^r} \qquad (\varphi \text{ is central})$$
$$= (\varphi s)^{p^r}$$

is valid. We conclude that $(\varphi s)^{p^r}$ is centralized by a, and by using theorem 2.4.8 part (i) is proven.

ad(ii): Let $r \in \mathbb{N}$ such that $o(\overline{\varphi^{H^X}}) = p^r$ is valid. Based on theorem 2.4.8 an element $\alpha \in C_{H^X}(\varphi^{p^r}) \setminus C_{H^X}(\varphi)$ exists. s is contained in the kernel of the action of δ, and thus $(\varphi s)^\alpha = \varphi^\alpha s \neq \varphi s$ is true. Hence, φs is not centralized by α. In addition,

$$((\varphi s)^{p^r})^\alpha \qquad (s \in ker\delta)$$
$$= (\varphi^{p^r} s^{p^r})^\alpha \qquad (\text{choice of } \alpha)$$
$$= \varphi^{p^r} (s^{p^r})^\alpha \qquad (s^{p^r} \in ker\delta)$$
$$= \varphi^{p^r} s^{p^r} \qquad (s \in ker\delta)$$
$$= (\varphi s)^{p^r}$$

is true. We deduce, that $(\varphi s)^{p^r}$ is centralized by α, and based on theorem 2.4.8 the first part of statement (ii) is proven. The add-on is a consequence of part (iii) of lemma 3.4.6.

ad(iii): Let $r \in \mathbb{N}$ such that $o(\varphi) = p^r$ is valid. Because of $\varphi \in Z(H^X)$ we deduce $p^r \leq exp(Z(H))$. s is contained in $ker\delta$, and thus $(\varphi s)^{p^r} = \varphi^{p^r} s^{p^r} = s^{p^r}$ is true. The element s^{p^r} is central in $H \wr_X S$, and therefor part (iii) is a consequence of theorem 2.4.8.

ad(iv): Let $r \in \mathbb{N}$ such that $p^r = max\{o(\varphi), o(\overline{s^S})\}$ is valid. Straightforward to prove is the identity $p^r \leq max\{exp(Z(H)), o(\overline{s^S})\}$. s is contained in the kernel of the action, and thus $(\varphi s)^{p^r} = \varphi^{p^r} s^{p^r} = s^{p^r}$ is true. We use theorem 2.4.8 to deduce the existence of an element $a \in C_S(s^{p^r}) \setminus C_S(s)$. $(\varphi s)^{p^r}$ is centralized by a. If a would centralize φs, then $\varphi s = \varphi^a s^a$ is valid. (H^X, S) is a semidirect decomposition of $H \wr_X S$, and thus s would be centralized by a which is a contradiction. We use theorem 2.4.8 to finalize the proof.◇

The results proven so far for the wreath product are used to deduce the following main theorem of this section:

3.5.11 Theorem (general wreath products)

Let p be a prime number, K a field, $char(K) = p$, H, S p-groups and S is acting by δ on a finite set X. Let $m := max\{o(\overline{s^S}) \mid s \in ker\delta \setminus Z(S)\}$. The exponent of the center of $rad(K(H \wr_X S))^*$ is exactly

$$max\{exp(Z(rad(KH)^*)), exp(Z(S) \cap ker\delta), m\}.$$

Proof. Based on proposition 2.4.7 the exponent of the center of $H \wr_X S$ and the maximum of the orders of the conjugacy class sums for all conjugacy classes of $H \wr_X S$ in $rad(K(H \wr_X S))^*$ are to be determined. The first part was calculated already within proposition 3.5.4, the orders of certain conjugacy class sums are determined in corollary 3.5.8 and in proposition 3.5.9. The missing orders of conjugacy class sums are bounded within lemma 3.5.10.◇

3.5.12 Corollary (wreath products and operation on cosets)

Let p be a prime number, K a field, $char(K) = p$, H, S p-groups, U a subgroup of S and S is acting by right multiplication on $S/_r U$. Let $m := max\{o(\overline{s^S}) \mid s \in core_S(U) \setminus Z(S)\}$. The exponent of the center of $rad(K(H \wr_{S/_r U} S))^*$ is exactly

$$max\{exp(Z(rad(KH)^*)), exp(Z(S) \cap core_S(U)), m\}.$$

Proof. The kernel of the action of S on $S/_r U$ is the core of U in S. Thus, the proof is a consequence of theorem 3.5.11.◇

3.5.13 Corollary (wreath products and faithful operation)

Let p be a prime number, K a field, $char(K) = p$, H, S p-groups and S is acting faithful by δ on a finite set X. The identity

$$exp(Z(rad(K(H \wr_X S))^*)) = exp(Z(rad(KH)^*))$$

is valid.

Proof. By definition of a faithful action the statement $ker\delta = \{1_S\}$ is true. Thus, the proof is a consequence of theorem 3.5.11.◇

3.5.14 Corollary (regular wreath products)

Let p be a prime number, K a field, $char(K) = p$, H, S p-groups. The identity

$$exp(Z(rad(K(H \wr S))^*)) = exp(Z(rad(KH)^*))$$

is valid.

Proof. The proof is a direct consequence of corollary 3.5.13.◇

If we choose $H = Z_p$ and S arbitrary, then we derive that every p-group S is an epimorphic image of a p-group W such that $Z(rad(KW)^*)$ is elementary-Abelian (see also [5] where a similar result is proven).

The exponent of the center for special classes of groups

3.5.15 Examples (Sylow subgroups of symmetric and linear groups)

(i) Let p be a prime number, K a field, $char(K) = p$ and $n \in \mathbb{N}$. Based on corollary 3.5.14 the identity $exp(Z(rad(K(Z_{p^n} \wr Z_p))^*)) = p^n$ is valid. Thus we can construct exponents of arbitrary powers of p by using non-Abelian p-groups.

(ii) Let p be a prime number, K a field, $char(K) = p$ and G a finite p-group. $Z(rad(K(Z_p \wr G))^*)$ is elementary-Abelian based on corollary 3.5.14.

(iii) Let p be a prime number, K a field, $char(K) = p$, $n \in \mathbb{N}$, p no divisor of $n!$ and P a p-Sylow subgroup of S_n.
Let us represent n as a p-adic number like $n = \sum_{i=0}^{r} a_i p^i$, and for every $i \in \underline{r}$ let P_i be a p-Sylow subgroup of S_{p^i}. We use the statements of the pages 176 and 177 in [56] to deduce that $P \cong_g P_1^{a_1} \times \cdots \times P_r^{a_r}$ is true. Based on a theorem of Kaloujnine for every $i \in \underline{r}$ an element $n_i \in \mathbb{N}$ exists such that $P_i \cong_g \underbrace{Z_p \wr \cdots \wr Z_p}_{n_i-\text{times}}$ is valid. By using the corollaries 3.5.14 and 3.4.8 we deduce that $Z(rad(KP)^*)$ is elementary-Abelian.

(iv) Let K be a finite field, $q := |K|$ and p a prime number such that $gcd(p, 2) = gcd(p, q)$ is valid. Let $e := min\{n \in \mathbb{N} \mid p \mid q^e - 1\}$ and $x, r \in \mathbb{N}$ such that $q^e - 1 = p^r x$ and $gcd(p, x) = 1$ are true. If P is a p-Sylow subgroup of

$GL(n, q)$ (linear group),
$C(2m, q)$ (symplectic group),
$U(n, q^2)$ (unitary group) or of
$O_D(n, q)$ (orthogonal group),

then we use [75] to deduce that an element $n \in \mathbb{N}$ exists such that $P \cong_g Z_{p^r} \wr \underbrace{Z_p \wr \cdots \wr Z_p}_{n-\text{times}}$ is true. Based on corollary 3.5.14 the exponent of $Z(rad(KP)^*)$ is exactly p^r. $\diamond$

3.5.16 Corollary (wreath product and conjugation)

Let p be a prime number, K a field, $char(K) = p$ and H, S p-groups. The following statements are valid:

(i) Let S be acting by conjugation on S. The exponent of the center of $rad(K(H \wr_S S))^*$ is exactly $max\{exp(Z(rad(KH)^*)), exp(Z(S))\}$.

(ii) Let $s \in S$, S acts by conjugation on s^S and $m := max\{o(\overline{s^S}) \mid s \in cores(C_S(s)) \setminus Z(S)\}$. The exponent of the center of $rad(K(H \wr_{s^S} S))^*$ is exactly $max\{exp(Z(rad(KH)^*)), exp(Z(S)), m\}$.

Proof. The kernel of the action of S on S resp. of S on s^S by conjugation is exactly $Z(S)$ resp. $cores(C_S(s))$. Thus, theorem 3.5.11 lets us finish the proof.◇

3.5.17 Remark

Let p be a prime number, K a field, $char(K) = p$ and H, S p-groups. S is acting trivial on a finite set X. In this case $H \wr_X S$ and $H^X \times S$ are $\mathcal{G}$-isomorphic, and based on corollary 3.4.8 the exponent of the center of $rad(H \wr_X S)^*$ is exactly $max\{exp(Z(rad(KH)^*)), exp(Z(rad(KS)^*))\}$.

We could also use theorem 3.5.11 to derive this result because the kernel of the trivial actions is S. But we have already used within the proof of theorem 3.5.11 the statement of corollary 3.4.8.◇

3.5.18 Corollary (bounds)

Let p be a prime number, K a field, $char(K) = p$, H, S p-groups and S acts on a finite set X. The following statements are valid:

(i) $exp(Z(rad(KH)^*)) \leq exp(Z(rad(K(H \wr_X S))^*))$

(ii) $exp(Z(rad(K(H\wr_X S))^*)) \leq max\{exp(Z(rad(KH)^*)), exp(Z(rad(KS)^*))\}$

(iii) If $exp(Z(rad(KH)^*)) \geq exp(Z(rad(KS)^*))$ is valid, then $exp(Z(rad(K(H \wr_X S))^*)) = exp(Z(rad(KH)^*))$ is true.

(iv) If $B_1, \ldots, B_r$ are the orbits of X under S, then $exp(Z(rad(K(H \wr_X S))^*)) \leq min\{exp(Z(rad(K(H \wr_{B_i} S))^*)) \mid i \in \underline{r}\}$ is valid.

Proof. All statements are a direct consequence of theorem 3.5.11.◇

3.5.19 Remark

(i) The lower resp. upper bound within part (i) resp. (ii) of corollary 3.5.18 is met for a faithful resp. trivial action (see corollary 3.5.13 and remark 3.5.17). The following examples illustrate that this is also valid for other actions:

(ii) Let U be a subgroup of order 2 in Z_8. Z_8 is acting by right multiplication on $Z_8/_r U$, and we focus on the wreath product $Z_2 \wr_{Z_8/_r U} Z_8$. As lower resp. upper bound we calculate 2 resp. 8, but the exact value is –

based on corollary 3.5.12 – the number 2.

(iii) Let $n \in \mathbb{N}_{\geq 4}$ and D_{2^n} the dihedral group of order 2^n $\mathcal{G}$-generated by two elements h, a. For these $o(h) = 2^{n-1}$, $o(a) = 2$ and $h^a = h^{-1}$ are valid. D_{2^n} is acting by right multiplication on $D_{2^n}/_r \langle h \rangle_{\mathcal{G}}$. We focus on the wreath product $Z_2 \wr_{D_{2^n}/_r \langle h \rangle} D_{2^n}$. As lower resp. upper bound we calculate 2 resp. (based on examples 3.1.2) 2^{n-2} which is – based on corollary 3.5.12 – the exact value.

(iv) Let U be a subgroup of order 4 in Z_8. Z_8 is acting by right multiplication on $Z_8/_r U$, and we focus on the wreath product $Z_2 \wr_{Z_8/_r U} Z_8$. As lower resp. upper bound we determine 2 resp. 8. The exact value is – based on corollary 3.5.12 – the number 4.◇

3.6 Special group extensions

We begin this chapter by presenting some aspects of the theory of O. Schreier[1] concerning group extensions.

3.6.1 Theorem

Let H, N groups and $N(\cdot; \cdot) : H \times H \longrightarrow N$ and $\alpha : H \longrightarrow Aut(N)$ functions possessing the following characteristics:

(a) $\forall h_1, h_2, h_3 \in H : N(h_1; h_2 h_3) N(h_2; h_3) = N(h_1 h_2; h_3)(N(h_1; h_2)\alpha(h_3))$

(b) $\forall n \in N, h_1, h_2 \in H : n(\alpha(h_1)\alpha(h_2)) = (n\alpha(h_1 h_2))^{N(h_1;h_2)}$

(c) $\forall h \in H : N(h; 1_H) = 1_N = N(1_H; h)$.

---

[1] Otto Schreier (3 March 1901 in Vienna, Austria to 2 June 1929 in Hamburg, Germany) was a Jewish-Austrian mathematician who made major contributions in combinatorial group theory and in the topology of Lie groups. He studied mathematics at the University of Vienna and obtained his doctorate in 1923, under the supervision of Philipp Furtwängler. He then moved to the University of Hamburg. Significance of the Artin-Schreier theorem according to Hans Zassenhaus: O. Schreier's and Artin's ingenious characterization of formally real fields as fields in which −1 is not the sum of squares and the ensuing deduction of the existence of an algebraic ordering of such fields started the discipline of real algebra. Really, Artin and his congenial friend and colleague Schreier set out on the daring and successful construction of a bridge between algebra and analysis. In the light of Artin-Schreier's theory the fundamental theorem of algebra truly is an algebraic theorem inasmuch as it states that irreducible polynomials over real closed fields only can be linear or quadratic. Results and concepts named after Otto Schreier are Nielsen-Schreier theorem, Schreier refinement theorem, Artin-Schreier theorem, Schreier's subgroup lemma, Schreier-Sims algorithm, Schreier coset graph, Schreier conjecture, Schreier domain.

If we define for all $h_1, h_2 \in H, n_1, n_2 \in N$
$(h_1; n_1) \cdot_{\alpha, N(\cdot;\cdot)} (h_2; n_2) := (h_1 h_2; N(h_1; h_2)(n_1 \alpha(h_2))n_2)$, then the following statements are true:

(i) $(H \times N; \cdot_{\alpha, N(\cdot;\cdot)})$ is a group.

(ii) The set $\{(1_H; n) \mid n \in N\}$ is a normal subgroup of $(H \times N; \cdot_{\alpha, N(\cdot;\cdot)})$ which is 9-isomorphic to N possessing a factor group 9-isomorphic to H.

(iii) $(1_H; 1_N)$ is the unit element in $(H \times N; \cdot_{\alpha, N(\cdot;\cdot)})$.

(iv) For all $h \in H, n \in N$ the identity $(h; n)^{-1} = (h^{-1}; (N(h^{-1}; h)n^{-1})\alpha(h)^{-1})$ is valid.

(v) $\alpha(1_H) = id_N$

(vi) If N is Abelian, then part (b) can be replaced by the statement (b') α is a 9-homomorphism.

Proof. see theorem 14.2 on page 87 in [26].⋄

3.6.2 Definition (group extensions)

Let H, N groups and $N(\cdot;\cdot) : H \times H \longrightarrow N$ and $\alpha : H \longrightarrow Aut(N)$ functions possessing the characteristics (a) to (c) within theorem 3.6.1. The group $(H \times N; \cdot_{\alpha, N(\cdot;\cdot)})$ is called the group extension of H and N based on the factor system $N(\cdot;\cdot)$ and the automorphism $\alpha(h), h \in H$.⋄

3.6.3 Theorem

Let G be a group, N be a normal subgroup of G and R a system of representatives of N in G containing the element 1_G. For all $r, s \in R$ let $N_R(r; s)$ resp. $t_{r,s}$ be the element of N resp. of R such that $rs = t_{r,s} N_R(r; s)$ is valid. Let $N_R(\cdot;\cdot) : G/N \times G/N \longrightarrow N$ defined by $(rN; sN) \mapsto N_R(r; s)$ and $\alpha_R : G/N \longrightarrow Aut(N)$ defined by $rN \mapsto (r\kappa)_{|N}$ for all $r, s \in R$. These functions possess the characteristics (a) to (c) of theorem 3.6.1.

Proof. The theorem is presented in theorem 14.1 on page 86 in [26].⋄

A remarkable connection between the theorems 3.6.1 and 3.6.3 is that all group extensions are presentable by the special ones used in theorem 3.6.3:

3.6.4 Proposition

Let G be a group, N a normal subgroup of G and R a system of representatives of N in G containing 1_G. The identity $G \cong_{\mathfrak{g}} (G/N \times N; \cdot_{\alpha_R, N_R(\cdot; \cdot)})$ is valid.

Proof. We define $\Phi : G/N \times N \longrightarrow G$ by $(rN; n) \mapsto rn$ for all $r \in R$ and $n \in N$. R is a system of representatives for N in G, and thus Φ is bijective. Let $r, s \in R$ and $n, m \in N$. We use $rs = t_{r,s} N_R(r; s)$ and obtain $((rN; n) \cdot_{\alpha_R, N_R(\cdot; \cdot)} (sN; m))\Phi = t_{r,s} N_R(r; s) n^s m$. By definition the identity $(rN; n)\Phi = rn$ und $(sN; m)\Phi = sm$ is true. In addition,
$rnsm = t_{r,s} N_R(r; s) n^s m \iff rns = t_{r,s} N_R(r; s) s^{-1} ns \iff rs = t_{r,s} N_R(r; s)$
is valid and the proof is finished.◇

If two group extensions E and $\hat{E}$ possess some special attributes, then we can prove that the exponents of $Z(rad(KE)^*)$ and $Z(rad(K\hat{E})^*)$ are identical. A starting point of this analysis is the following lemma:

3.6.5 Lemma

Let H, N and $\hat{H}, \hat{N}$ groups possessing factor systems $N(\cdot; \cdot)$ and $\hat{N}(\cdot; \cdot)$ and automorphism $\alpha(h), h \in H$ and $\hat{\alpha}(\hat{h}), \hat{h} \in \hat{H}$. Let $\varphi : N \longrightarrow \hat{N}$ and $\psi : H \longrightarrow \hat{H}$ two $\mathfrak{g}$-isomorphism and $\phi : Aut(N) \longrightarrow Aut(\hat{N})$, $\beta \mapsto \beta^\varphi$. The following conditions are assumed:

(i) $\psi \hat{\alpha} = \alpha \phi$
   (The actions from H on N and from $\hat{H}$ on $\hat{N}$ are equivalent.)

(ii) $N(\cdot; \cdot)$ and $\hat{N}(\cdot; \cdot)$ are symmetric.
   (This statement is e.g. true if N and $\hat{N}$ are Abelian.)

(iii) For all $x \in im\, N(\cdot; \cdot)$ resp. $\hat{x} \in im\, \hat{N}(\cdot; \cdot)$ and all $h \in H$ resp. $\hat{h} \in \hat{H}$ the identity $x\alpha(h) = x$ resp. $\hat{x}\hat{\alpha}(\hat{h}) = \hat{x}$ is valid.

Let $h \in H$, $n \in N$ and $s \in \mathbb{N}$. The centralizer of $(h; n)^s$ in $(H \times N; \cdot_{\alpha, N(\cdot; \cdot)})$ possesses the same order as the one of $(h\psi; n\varphi)^s$ in $(\hat{H} \times \hat{N}; \cdot_{\hat{\alpha}, \hat{N}(\cdot; \cdot)})$.

Proof. An induction argument on s leads to

(1) $(h; n)^s = (h^s; N(h; h^{s-1})(n\alpha(h^{s-1})) \ldots N(h; h)(n\alpha(h))n)$.

Let $(h_1; n_1) \in H \times N$. This element centralizes $(h; n)^s$ because of (ii) and (iii) if and only if $h_1 h = h h_1$ and

(2) $(n_1 \alpha(h^s))(n\alpha(h^{s-1})) \ldots (n\alpha(h))n = ((n\alpha(h^{s-1})) \ldots (n\alpha(h))n)\alpha(h_1) n_1$

are valid. For all $i \in \underline{s}$ we use (i) to deduce

(3) $(n\varphi)(\hat{\alpha}(h^i\psi)) = (n\alpha(h^i))\varphi$.

By using (1) and (3) we obtain

(4) $(h\psi; n\varphi)^s = (h^s\psi; \hat{N}(h\psi; h\psi^{s-1})(n\alpha(h^{s-1}))\varphi \ldots \hat{N}(h\psi; h\psi)(n\alpha(h))\varphi n\varphi)$.

Now we use (2), (4), (ii) and (iii) and prove that $(h_1\psi; n_1\varphi)$ is centralized by $(h\psi; n\varphi)^s$ if and only if $(h_1; n_1)$ centralizes the element $(h; n)^s$. The function $H \times N \longrightarrow \hat{H} \times \hat{N}$, $(a; b) \mapsto (a\psi; b\varphi)$ is bijective, and thus the lemma is proven.⋄

3.6.6 Theorem (group extensions)

Let p be a prime number, K a field, $char(K) = p$, $G, \hat{G}$ p-groups, N resp. $\hat{N}$ a normal subgroup of G resp. of $\hat{G}$ and R resp. $\hat{R}$ a system of representatives of N in G resp. of $\hat{N}$ in $\hat{G}$ containing $1_G \in R$ resp. $1_{\hat{G}} \in \hat{R}$. Let $\varphi : N \longrightarrow \hat{N}$ and $\quad : G/N \longrightarrow \hat{G}/\hat{N}$ $\mathcal{G}$-isomorphism and $\phi : Aut(N) \longrightarrow Aut(\hat{N})$, $\beta \mapsto \beta^\varphi$. The functions $N_R(\cdot; \cdot)$, $N_{\hat{R}}(\cdot; \cdot)$, α_R and $\alpha_{\hat{R}}$ possess the following attributes:

(i) $\psi \alpha_{\hat{R}} = \alpha_R \phi$

(ii) $N_R(\cdot; \cdot)$ and $N_{\hat{R}}(\cdot; \cdot)$ are symmetric.

(iii) For all $x \in im\, N_R(\cdot; \cdot)$ resp. $\hat{x} \in im\, \hat{N}R(\cdot; \cdot)$ and all $r \in R$ resp. $\hat{r} \in \hat{R}$ the identity $x^r = x$ resp. $\hat{x}^{\hat{r}} = \hat{x}$ is valid.
(This statement is true, if e.g. $Im\, N_R(\cdot; \cdot) \subseteq Z(G)$ and $Im\, N_{\hat{R}}(\cdot; \cdot) \subseteq Z(\hat{G})$ are valid.)

The maxima of the sets $\{o(\overline{g^G}) \mid g \in G \setminus Z(G)\}$ and $\{o(\overline{\hat{g}^{\hat{G}}}) \mid \hat{g} \in \hat{G} \setminus Z(\hat{G})\}$ are identical. If, in addition, the centers of G and $\hat{G}$ are cyclic, then the exponents of $Z(rad(KG)^*)$ and $Z(rad(K\hat{G})^*)$ are identical.

Proof. The first part is a consequence of theorem 2.4.8, of proposition 3.6.4 and of lemma 3.6.5. Based on lemma 3.6.5 the centers of G and $\hat{G}$ are of the same order. If both are cyclic, then both are $\mathcal{G}$-isomorphic. We finish the proof by using proposition 2.4.7.⋄

3.6.7 Example

Let K be a field, $char(K) = 2$ and $n \in \mathbb{N}_{\geq 3}$. Let $G := D_{2^n}$ be the dihedral group of order 2^n, and let $h, a \in G$ such that $G = \langle h, a \rangle_{\mathcal{G}}$, $o(h) = 2^{n-1}$, $o(a) = 2$ and $h^a = h^{-1}$ are valid. Let $\hat{G} := Q_{2^n}$ be the quaternion

The exponent of the center for special classes of groups 93

group of order 2^n, and let $x, y \in \hat{G}$ such that $\hat{G} = \langle x, y \rangle_{\mathcal{G}}$, $o(y) = 2^{n-1}$, $x^2 = y^{2^{n-2}}$ and $y^x = y^{-1}$ are true. Within example 3.1.2 we have proven that $exp(Z(rad(KD_{2^n})^*)) = exp(Z(rad(KQ_{2^n})^*)) (= 2^{n-2})$ is valid. Let $N := \langle h \rangle_{\mathcal{G}}$ and $\hat{N} := \langle y \rangle_{\mathcal{G}}$. $R := \{1_G, a\}$ resp. $\hat{R} := \{1_{\hat{G}}, y\}$ is a system of representatives for N in G resp. for $\hat{N}$ in $\hat{G}$. Let $\varphi : N \longrightarrow \hat{N}$ defined by $a^i \mapsto y^i$ for all $i \in \underline{2^{n-1}}$ and $\psi : G/N \longrightarrow \hat{G}/\hat{N}$ defined by $N \mapsto \hat{N}$ und $aN \mapsto y\hat{N}$. The functions φ and ψ are $\mathcal{G}$-isomorphism. The attributes (i) and (ii) and the add-on of theorem 3.6.6 are straightforward to verify. Now we prove that also part (iii) of this theorem is valid. For the corresponding factor systems we calculate:
$N_R(1_G; 1_G) = N_R(1_G; a) = N_R(a; 1_G) = N_R(a; a) = 1_G \in Z(G)$,
$N_{\hat{R}}(1_{\hat{G}}; 1_{\hat{G}}) = N_{\hat{R}}(1_{\hat{G}}; y) = N_{\hat{R}}(y; 1_{\hat{G}}) = 1_{\hat{G}} \in Z(\hat{G})$ and
$N_{\hat{R}}(y; y) = y^2 \in Z(\hat{G}).\diamond$

3.7 Remarks to the bounds within section 2.5

3.7.1 Examples for 2.5.6

Let p be a prime number, K a field, $char(K) = p$, $n \in \mathbb{N}_{\geq 4}$ and $G = \langle h, a \rangle_{\mathcal{G}}$ a p-group such that $o(h) = p^n$, $o(a) = p$, $|G| = p^{n+1}$ and $h^a = h^{1+p^{n-1}}$ are valid. Because of $(h^p)^a = (h^a)^p = (h^{1+p^{n-1}})^p = h^p$ we obtain $Z(G) = \langle h^p \rangle_{\mathcal{G}}$. We use theorem 2.4.8 and derive $o(\overline{h^G}) = p < p^2$. In addition, G is the only subgroup of G in which h is not central.$\diamond$

3.7.2 Examples for 2.5.8

Let K be a field, $|K| = 2$ and $G := 1_{K^{n \times n}} + su(4, K)$. Because of theorem 3.2.2.5 the center of $rad(KG)^*$ is elementary-Abelian. Now we prove that $x, y \in G$ exist such that $[x^2, y] = 1_G$ and $[x, y^2] \neq 1_G$ are valid.

Let $x := \begin{pmatrix} 1_K & 0_K & 0_K & 0_K \\ 1_K & 1_K & 0_K & 0_K \\ 0_K & 0_K & 1_K & 0_K \\ 0_K & 0_K & 0_K & 1_K \end{pmatrix}$ and $y := \begin{pmatrix} 1_K & 0_K & 0_K & 0_K \\ 0_K & 1_K & 0_K & 0_K \\ 0_K & 1_K & 1_K & 0_K \\ 0_K & 0_K & 1_K & 1_K \end{pmatrix}$.

We calculate $x^2 = 1_G$ and $y^2 = \begin{pmatrix} 1_K & 0_K & 0_K & 0_K \\ 0_K & 1_K & 0_K & 0_K \\ 0_K & 0_K & 1_K & 0_K \\ 0_K & 1_K & 1_K & 1_K \end{pmatrix}$.

In addition, $[x^2, y] = 1_G$ is valid and because of

$y^2 x = \begin{pmatrix} 1_K & 0_K & 0_K & 0_K \\ 1_K & 1_K & 0_K & 0_K \\ 0_K & 0_K & 1_K & 0_K \\ 1_K & 1_K & 1_K & 1_K \end{pmatrix}$ and $xy^2 = \begin{pmatrix} 1_K & 0_K & 0_K & 0_K \\ 1_K & 1_K & 0_K & 0_K \\ 0_K & 0_K & 1_K & 0_K \\ 0_K & 1_K & 1_K & 1_K \end{pmatrix}$

we deduce $[y^2, x] \neq 1_G$. We remark that for all $x \in G$ the identity $x^4 = 1_G$ is true. The minimal $n \in \mathbb{N}$ possessing the characteristic that for all $x, y \in G$ the elements x^{2^n} and y commute if and only if the elements x and y^{2^n} commute is 2 and not 1.◇

3.7.3 Examples for 2.5.14

(1) Let K be a field, $char(K) = 2$, $n \in \mathbb{N}_{\geq 4}$ and G the dihedral group of order 2^n. Elements $a, b \in G$ exist such that $G = \langle a, b \rangle_{\mathcal{G}}$, $o(a) = 2^{n-1}$, $o(b) = 2$ and $a^b = a^{-1}$ are valid. $\langle a^{(2^{n-2})} \rangle_{\mathcal{G}}$ is the center and $\langle a^2 \rangle_{\mathcal{G}}$ is the derived subgroup of G.

(i) We use example 3.1.2 and deduce
$exp(Z(rad(KG)^*)) = 2^{n-2}$,
$2^{n-2} = 2^{n-3} \cdot 2 = exp(Z(rad(K(G/Z(G)))^*)) \cdot exp(rad(KZ(G))^*)$ and
$2^{n-1} = 2^{n-1} \cdot 2 = exp(rad(KG')^*) \cdot exp(rad(K(G/G'))^*)$.

(ii) We calculate $exp(Z(rad(KZ(G))^*)) = 2 < 2^{n-2}$,
$exp(Z(rad(KG')^*)) = 2^{n-2}$ and $exp(Z(rad(K\langle h \rangle_{\mathcal{G}})^*)) = 2^{n-1} > 2^{n-2}$.

(iii) Based on corollary 3.4.8 we obtain
$exp(Z(rad(K(G \times G))^*)) = 2^{n-2} = exp(Z(rad(K((G \times G)/(G \times \{1_G\})))^*))$
and $exp(Z(rad(K(G/Z(G)))^*)) = 2^{n-3} < 2^{n-2}$.

(2) Let p be a prime number, $p \neq 2$, K a field, $char(K) = p$ and G the semi-direct product of $\langle a \rangle_{\mathcal{G}}$ and $\langle b \rangle_{\mathcal{G}}$ such that $o(a) = p^3$, $o(b) = p^2$ and $a^b = a^{1+p}$ are valid. Let $\alpha : \langle a \rangle_{\mathcal{G}} \longrightarrow \langle a \rangle_{\mathcal{G}}$ defined by $a^i \mapsto (a^i)^{p+1}$. The function α is a $\mathcal{G}$-automorphism of $\langle a \rangle_{\mathcal{G}}$ of order $o(\alpha) = p^2$. G' is cyclic and $p \neq 2$ is valid. Thus, G is a regular p-group (see theorem 10.2 on page 322 in [26]). We use corollary 3.3.3 and deduce $exp(Z(rad(KG)^*)) = exp(Z(G))$. Now we prove that $exp(Z(G)) = p$ and $exp(Z(rad(KG)^*)) = p < p^2 = exp(Z(rad(K(G/\langle a \rangle_{\mathcal{G}}))^*))$ are valid. Let $i \in \underline{p}^3$ and $j \in \underline{p}^2$. We calculate:
$(a^i b^j)^b = a^i b^j \iff (a^b)^i = a^i \iff a^{i(p+1)} = a^i \iff p^2 \mid i$.
In addition, $(a^i b^j)^a = a^i b^j$ is valid if and only if $a \in C_G(\langle b^j \rangle_{\mathcal{G}})$ is true. If $\langle b^j \rangle_{\mathcal{G}} = \langle b \rangle_{\mathcal{G}}$ would be valid, then $a = a^b = a^{1+p}$ and $a^p = 1_G$ would be true which is a contradiction to $o(a) = p^3$. If $\langle b^j \rangle_{\mathcal{G}} = \langle b^p \rangle_{\mathcal{G}}$ would be true, then $a^{b^p} = a$ would be valid. Thus, $a(\alpha^p) = a$ and $\alpha^p = id_{\langle a \rangle_{\mathcal{G}}}$ would be true which is a contradiction to $o(\alpha) = p^2$. Therefor, we deduce $\langle b^j \rangle_{\mathcal{G}} = \{1_G\}$ and $Z(G) = \langle a^{p^2} \rangle_{\mathcal{G}} \cong_{\mathcal{G}} Z_p$.◇

At the end of this chapter we summarize the main results on exponents of this chapter – for $n \in \mathbb{N}$, $p \in \mathbb{P}$, q a power of p, finite p-groups G, H, $m \in G$, and a field K of characteristic p – within the following table:

The exponent of the center for special classes of groups

group	field	exponent $Z(rad(KG)^\star)$
G Abelian	K	$exp(Z(G))$
D_{2^n}	$p=2$	2^{n-2}
SD_{2^n}	$p=2$	2^{n-2}
Q_{2^n}	$p=2$	2^{n-2}
G	K	$\geq max\{p, exp(Z(G))\}$
G	K	$\leq min\{\frac{|G|}{p^2}, exp(G/Z(G))\}$
$\langle m \rangle_\mathcal{G} \trianglelefteq_{max} G$	K	$\frac{|G|}{p^2}$
$Z(G) \cong Z_{\frac{|G|}{p^2}}$	K	$\frac{|G|}{p^2}$
$\forall g \in G \setminus Z(G): C_G(g) < C_G(g^p)$	K	$exp(Z(G))$
$\forall g \in G \setminus Z(G): C_G(g) < C_G(g^p)$ and $Z(G)$ elementary-p-Abelian	K	p
$G \in Syl_p(GL(n, GF(q)))$	K	p
$G \in Syl_p(PGL(n, GF(q)))$	K	p
$G \in Syl_p(SL(n, GF(q)))$	K	p
$G \in Syl_p(PSL(n, GF(q)))$	K	p
$cl(G) \leq 2$	K	$exp(Z(G))$
$\mid G' \mid \leq p$	K	$exp(Z(G))$
$G^p \leq Z(G)$	K	$exp(Z(G))$
$\Phi(G) \leq Z(G)$	K	$exp(Z(G))$
G special	K	p
G extra-special	K	p
G regular	K	$exp(Z(G))$
G minimal non-Abelian	K	$exp(Z(G))$
$G \curlyvee H$	K	$max\{exp(Z(rad(KG_1)^\star)), exp(Z(rad(KG_2)^\star))\}$
$G \times H$	K	$max\{exp(Z(rad(KG_1)^\star)), exp(Z(rad(KG_2)^\star))\}$
$G \cong Q_8 \times (Z_2)^n$, Hamiltonian	K	2
$G \wr H$	K	$exp(Z(rad(KG)^\star))$
$Z_{p^n} \wr Z_{p^n}$	K	p^n
$Z_p \wr G$ (G is epimorphic image)	K	p
$H \in Syl_p(S_{p^n})$	K	p
$H \in Syl_p(S_{|G|})$ (G is a subgroup)	K	p

We remark that some more results on the exponents will be proven within the next chapters for isoclinic groups, semi-extra-special groups, VZ-groups, Camina groups, generalized Camina groups and on the iteration by $G = G_0$, $1 + rad(KG) = G_1$, $1 + rad(KG_1) = G_2$ and so on.

3.8 Open topics and exercises

Open-ended questions 3 *(i) In what way can the exponent of the center of the radical be determined for arbitrary group extensions?*

(ii) In what way can the exponent of the center of the radical be determined for the subdirect product of quotient groups, meta-Abelian groups, powerful (embedded) groups and groups of maximal class?

(iii) In what way can the structure of the center of the radical be determined for groups and group classes within this chapter? Some answers are presented within chapter 4.

(iv) Determine the conditions such that the maximal (e.g. for trivial action) and the minimal (e.g. for faithful action) possible exponent of the center of the radical is met for arbitrary wreath products!

Excercise 93 Let p be a prime number, K a field, $char(K) = p$ and G a finite p-group. Prove that $exp(G) = exp(N_{1+rad(KG)}(G))$ is valid.

Excercise 94 Let p be a prime number, K a field, $char(K) = p$, G a finite p-group, N a normal subgroup of G and $g \in G \setminus Z(G)$. Does a connection between the orders of $\overline{g^G}$ and $\overline{(gN)^{G/N}}$ exist?

Excercise 95 Let p be a prime number, K a field, $char(K) = p$, G a finite p-group, N a normal subgroup of G, U a subgroup of G and $g \in G \setminus Z(G)$. Prove or disprove the following statements:

(i) $g^G \subseteq gG'$

(ii) If $g^G = gG'$ is valid, then $\overline{g^G}$ is of order p.

(iii) If $g^G = gN$ is true, then $\overline{g^G}$ is of order p.

(iv) Let $h \in G$ such that $gU = Uh$ is valid. $\overline{gU}$ is of order p.

(v) Let $h \in G$ such that $gU = Uh$ is true. $gU = Uh = hU = Ug$ is valid.

Excercise 96 Try to enhance proposition 3.3.3 by using commutators of the form $[a^{p^n}, b]$ instead of the case $n = 1$!

Excercise 97 Let p be a prime number, K a field, $char(K) = p$ and G, H finite p-groups. True or false:

(i) $Z(1+rad(KG))$ and $Z(1+rad(KH))$ are of the same exponent if and only if G, H are of the same class of nilpotency.

(ii) $Z(1+rad(KG))$ and $Z(1+rad(KH))$ are of the same exponent if and only if G, H are of the same coclass of nilpotency.

(iii) $Z(1+rad(KG))$ and $Z(1+rad(KH))$ are of the same exponent if and only if G, H are of the same exponent.

(iv) $Z(1+rad(KG))$ and $Z(1+rad(KH))$ are of the same exponent if and only if $Z(G), Z(H)$ are of the same exponent.

Excercise 98 Let K a field, $char(K) = 2$ and G a finite metacyclic 2-group. Use [67] to describe the exponent the center of $1 + rad(KG)$ for all isomorphism types of G. Compare the results to metacyclic p-groups for odd p.

Excercise 99 Let p be a prime number, K a field, $char(K) = p$, G a finite p-group and $g \in G$. Prove $o(\overline{g^G}) = o(\overline{(g^{-1})^G})$.

Excercise 100 Let $p \neq 2$ be a prime number, K a field, $char(K) = p$, G a finite p-group of maximal class. Do a research in the literature that an element $g \in G$ exists such that $C_G(g)$ is of order p^2 and the center of G is of order p. Prove that $o(\overline{g^G}) = p$ is valid.

Excercise 101 Let K be a field of characteristic 2 and G a finite 2-group of maximal class. Determine the exponent of $Z(1 + rad(KG))$!

Excercise 102 Let p be a prime number, K a field, $char(K) = p$, G a finite p-group and $g, h \in G$. Prove or disprove $o(\overline{(gh)^G}) = o(\overline{g^G}) \cdot o(\overline{h^G})$.

Excercise 103 Apply proposition 3.1.1 to relevant subgroups of D_{16}, Q_{16} and SD_{16}. What is the consequence of proposition 3.1.1 for a group of order p^3 (p a prime number)?

Excercise 104 Within proposition 3.1.1 prove the statements for the semi-dihedral and quaternion groups in details.

Excercise 105 Let p be a prime number. Check whether theorem 3.1.6 can be applied to the group G:

(i) G is a regular p-group.

(ii) G is a minimal non-Abelian p-group.

(iii) G is a special p-group.

(iv) G is an extra-special p-group.

(v) $G \in \{D_{16}, Q_{16}, SD_{16}\}$

(vi) G is a 3-Sylow subgroup of S_5.

(vii) G is of order p^3.

(viii) G is the direct product of D_8 with Z_{32}.

(ix) G is a metacyclic p-group for odd p.

(x) G is a metacyclic 2-group.

(xi) G' is cyclic.

If the theorem can be applied to the group, then derive the corresponding result for it. In the other case present a counter-example and check the possibility to change the parameters such that the theorem can be used.

Excercise 106 Let p be a prime number, K a field, $char(K) = p$ and G_1, G_2 p-groups such that the exponent of the center of $rad(KG)$ with respect to $*$ is as large resp. small as possible for both groups. Is this characteristic also valid for their direct product, for their regular wreath product and for their central product? Is this characteristic also true for every subgroup, normal subgroup and quotient group of G_1?

Excercise 107 Why is every p-group isomorphic to a subgroup of a p-Sylow subgroup of a general linear group?

Excercise 108 Prove remark 3.2.2.1 in details (if needed by doing a research in the literature).

Excercise 109 We focus on a 31-Sylow subgroup P of $GL(17, 31^5)$. In what way can we present this Sylow subgroup? What is the exponent of the center of $rad(GF(31^5)P)^*$? What is the order of this center? How can we describe the structure of this center with respect to $*$?

Excercise 110 Try to extend exercise 109 to arbitrary prime numbers and dimensions.

Excercise 111 Let p be a prime number, K a field, $char(K) = p$ and G a finite p-group. Determine the exponent of the center of $rad(KG)^*$ for the following groups G:

(i) G is a 2-group such that G' is of order 2.

(ii) G is a 3-group such that $G^3 \leq Z(G)$ is valid.

(iii) G is a 5-group possessing a central Frattini subgroup.

(iv) G is a minimal non-Abelian 7-group.

(v) G is a meta-cyclic 11-group.

(vi) G is an extra-special 2-group.

(vii) G is the direct product of D_{18} with a group of order 8.

(viii) G is a Hamiltonian group of order 2^5 or 2^6.

(ix) G is a regular 17-group of order 17^{35}.

(x) G is a special 31-group of order 31^{17}.

(xi) G is a minimal non-Abelian 5-group of type 2 with $r = 6$ and $s = 4$.

(xii) G is a minimal non-Abelian 7-group of type 3 with $r = 8$ and $s = 9$.

Excercise 112 *Prove proposition 3.2.2.3 in details.*

Excercise 113 *Prove proposition 3.2.2.4 in details.*

Excercise 114 *Prove theorem 3.2.2.5 in details.*

Excercise 115 *Within remark 3.3.6 prove statement (1) in details.*

Excercise 116 *Prove theorem 3.4.2 in details (if needed by doing a research in the literature).*

Excercise 117 *For all pairs of the following groups analyze the possibility of constructing a central product P (for the same prime number). Consider the case of two identical groups, too.*

(i) D_8

(ii) Q_8

(iii) SD_8

(iv) Z_{16}

(v) Z_{3^5}

(vi) G is a group of order 3^3.

What is the exponent of the center of the radical of $GF(p)P$ with respect to the composition $*$?

Excercise 118 *Prove remark 3.4.3 in details.*

Excercise 119 *Prove proposition 3.5.1 in details.*

Excercise 120 *What is the effect within definition 3.5.2 by using the left regular representation?*

Excercise 121 *For every prime number p and p-Sylow subgroup P of S_4, A_4, S_5, A_5 and S_6 determine the exponent of the center of $rad(GF(p)P)^*$.*

Excercise 122 *Prove theorem 3.6.1 in details (if needed by doing a research in the literature).*

Excercise 123 *Prove the statements within definition 3.6.2 in details (if needed by doing a research in the literature).*

Excercise 124 *Execute example 3.6.7 for quaternion and semi-dihedral groups.*

Excercise 125 *Let p be a prime number, K a field and $char(K) = p$. For the following groups G describe the determination of the exponent of the center of $rad(KG)^*$:*

(i) G is an iterated regular wreath product of cyclic p-groups.

(ii) G is an iterated regular wreath product of Abelian p-groups.

(iii) G is an iterated regular wreath product of p-groups.

Chapter 4

The invariants of the center

4.1 A direct decomposition

4.1.1 Example

Let $G := D_{16}$, $h, a \in G$ such that $G = <h, a>_\mathfrak{g}$, $o(h) = 8$, $o(a) = 2$ and $h^a = h^{-1}$ are valid. The conjugacy classes of G are $\{h, h^7\}$, $\{h^2, h^6\}$, $\{h^3, h^5\}$, $\{1\}$, $\{h^4\}$, $\{a, h^2a, h^4a, h^6a\}$ and $\{ha, h^3a, h^5a, h^7a\}$. Let K be a field and $char(K) = 2$. We visualize the multiplication for the commutative K-algebra $Z(rad(KG))$ (see proposition 1.3.11):

	$\overline{h^G}$	$\overline{(h^2)^G}$	$\overline{(h^3)^G}$	$\overline{a^G}$	$\overline{(ah)^G}$	$h^4 - 1_G$
$\overline{h^G}$	$\overline{(h^2)^G}$	$\overline{h^G} + \overline{(h^3)^G}$	$\overline{(h^2)^G}$	0_{KG}	0_{KG}	$\overline{h^G} + \overline{(h^3)^G}$
$\overline{(h^2)^G}$		0_{KG}	$\overline{(h^2)^G} + \overline{(h^3)^G}$	0_{KG}	0_{KG}	0_{KG}
$\overline{(h^3)^G}$			$\overline{(h^2)^G}$	0_{KG}	0_{KG}	$\overline{h^G} + \overline{(h^3)^G}$
$\overline{a^G}$				0_{KG}	0_{KG}	0_{KG}
$\overline{(ah)^G}$					0_{KG}	0_{KG}
$h^4 - 1_G$						0_{KG}

Within this example the following statements are true:

(1) The K-subspace $\langle \{\overline{g^G} \mid g \in G \setminus Z(G)\} \rangle_K$ is a K-ideal of $Z(rad(KG))$.

(2) A refinement of (1) is the statement that the two K-subspaces $\langle \overline{h^G}, \overline{(h^2)^G}, \overline{(h^3)^G} \rangle_K$ and $\langle \overline{a^G}, \overline{(ha)^G} \rangle_K$ are K-ideals of $Z(rad(KG))$. The classes $h^G, (h^2)^G, (h^3)^G$

are of length 2, and the classes $a^G, (ha)^G$ are of length 4.

We will prove statement (1) but disprove that K-spaces of conjugacy class sums of a given size are K-ideals of $Z(rad(KG))$.◇

4.1.2 Remark

Let G be a finite group, U a subgroup of G, $g \in G$ and $c \in C_G(U)$. If $g^{u_1}, \ldots, g^{u_r}$ are the conjugates of g under U, then the following statements are valid:

(i) $cg^{u_1}, \ldots, cg^{u_r}$ are the conjugates of cg under U.

(ii) $g^{u_1}c, \ldots, g^{u_r}c$ are the conjugates of gc under U.

(iii) $(g^{-1})^{u_1}, \ldots, (g^{-1})^{u_r}$ are the conjugates of g^{-1} under U.◇

4.1.3 Lemma

Let K be a field, G a finite group, U a subgroup of G, $\mathcal{B}$ the set of U-orbits of G by conjugation of G and $C, D \in \mathcal{B}$. For all $B \in \mathcal{B}$ let $k_B \in K$ such that $\overline{C} \cdot \overline{D} = \sum_{B \in \mathcal{B}} k_B \overline{B}$ is true (see proposition 1.3.14). If $z \in C_G(U)$ and $(c; d) \in C \times D$ is a pair such that $cd = z$ is valid, then $k_{\{z\}} = \mid C \mid_K = \mid D \mid_K$ is true.

Proof. Let $c^{u_1}, \ldots, c^{u_r}$ be the conjugates of c under U. We use $d = c^{-1}z$ and remark 4.1.2 to deduce that $d^{u_1}, \ldots, d^{u_r}$ are the conjugates of d under U. Because of $cd = z \in C_G(U)$ for all $i \in \underline{r}$ the equation $z = z^{u_i} = (cd)^{u_i} = c^{u_i}d^{u_i}$ is true. Let $i, j \in \underline{r}$ such that $c^{u_i}d^{u_j} = z$ is valid. $c^{u_i}d^{u_i} = z$ is true and we conclude $d^{u_i} = d^{u_j}$. Thus, $i = j$ is valid and the proof is finished.◇

4.1.4 Theorem

Let p be a prime number, K a field, $char(K) = p$, G a p-group, U a subgroup of G and $\mathcal{B}$ the set of U-orbits of G by conjugation. The following statements are valid:

(i) $(\langle\{\overline{B} \mid B \in \mathcal{B}, \mid B \mid \neq 1\}\rangle_K, KC_G(U))$ is a semidirect decomposition of the K-algebra $C_{KG}(U)$.

(ii) $(\langle\{\overline{B} \mid B \in \mathcal{B}, \mid B \mid \neq 1\}\rangle_K, rad(KC_G(U)))$ is a semidirect decomposition of the K-algebra $C_{rad(KG)}(U)$.

(iii) $(\langle\{\overline{B} \mid B \in \mathcal{B}, \mid B \mid \neq 1\}\rangle_K^*, rad(KC_G(U))^*)$ is a semidirect decomposition of the group $C_{rad(KG)^*}(U-1)$.

The invariants of the center 103

(iv) $(1_G + \langle\{\overline{B} \mid B \in \mathcal{B}, \mid B \mid \neq 1\}\rangle_K, 1_G + rad(KC_G(U)))$ is a semidirect decomposition of the group $C_{1_G+rad(KG)}(U)$.

Proof. ad(i): Part (i) is a consequence of proposition 1.3.14, lemma 4.1.3 and parts (i) and (ii) of remark 4.1.2.

ad(ii): We use (i) and corollary 1.3.16 to obtain (ii).

ad(iii): This statement is a consequence of (ii) and corollary 1.1.8.

ad(iv): We use (iii) and corollary 1.1.8 to deduce part (iv).⋄

By specializing $U = G$ we obtain the following corollary:

4.1.5 Corollary (decomposition of the center)

Let p be a prime number, K a field, $char(K) = p$ and G a p-group. The following statements are valid:

(i) $(\langle\{\overline{C} \mid C \in \mathcal{K}(G), \mid C \mid \neq 1\}\rangle_K, KZ(G))$ is a semidirect decomposition of the K-algebra $Z(KG)$.

(ii) $(\langle\{\overline{C} \mid C \in \mathcal{K}(G), \mid C \mid \neq 1\}\rangle_K, rad(KZ(G)))$ is a semidirect decomposition of the K-algebra $Z(rad(KG))$.

(iii) $(\langle\{\overline{C} \mid C \in \mathcal{K}(G), \mid C \mid \neq 1\}\rangle_K^*, rad(KZ(G))^*)$ is a direct decomposition of the group $Z(rad(KG)^*)$.

(iv) $(1_G + \langle\{\overline{C} \mid C \in \mathcal{K}(G), \mid C \mid \neq 1\}\rangle_K, 1_G + rad(KZ(G)))$ is a direct decomposition of the group $Z(1_G + rad(KG))$.⋄

4.1.6 Definition (ideal of non-central class sums)

Let K be a field and G a finite group. We define

$$\overline{\mathcal{K}(G)} := \langle\{\overline{C} \mid C \in \mathcal{K}(G), \mid C \mid \neq 1\}\rangle_K.$$

This set is – in the case of $char(K) = p > 0$ and of a p-group G – based on part (i) of corollary 4.1.5 a K-ideal of $Z(KG)$ and, in particular, a K-subalgebra of KG.⋄

4.1.7 Example

(i) Let p be a prime number and G a non-Abelian group of order p^3. G is extra-special, and thus for all $g \in G \setminus Z(G)$ the identity $\mid g^G \mid = p$ is true. If $n \in \mathbb{N}$ is valid, then for all $g \in G \setminus Z(G)$ the statement $\mid (g, \ldots, g)^{G^n} \mid = p^n$ is true.

(ii) Let K be a field, A, B groups and $(a; b) \in A \times B$ such that $a \notin Z(A)$ and $b \notin Z(B)$ are true. A straightforward calculation lets us deduce that $\overline{(a; 1_B)^{A \times B}} \cdot \overline{(1_A; b)^{A \times B}} = \overline{(a; b)^{A \times B}}$ is valid. The length of the conjugacy class $(a; b)^{A \times B}$ is exactly the product of the length of the conjugacy classes $(a; 1_B)^{A \times B}$ and $(1_A; b)^{A \times B}$.

(iii) Based on (i) and (ii) we deduce that a K-space of conjugacy class sums of a fixed length are not closed in general under $\cdot$ and $*$.◇

4.2 Commutative group algebras

4.2.1 The invariants

4.2.1.1 Definition (n-fold copies)

Let G be a group and $n \in \mathbb{N}$. We define $nG := G^n = \underbrace{G \times \cdots \times G}_{n-times}$.◇

The next proposition is straightforward to be proven but very essential for determining the invariants of an Abelian p-group:

4.2.1.2 Proposition

Let p be a prime number, $e \in \mathbb{N}$, $n_i \in \mathbb{N}$ for all $i \in \underline{e}$ and G a group $\mathcal{G}$-isomorphic to $n_1 Z_p \times \cdots \times n_e Z_{p^e}$. For all $i \in \underline{e-1} \cup \{0\}$ the factor group $G^{p^i}/G^{p^{i+1}}$ is $\mathcal{G}$-isomorphic to $(n_{i+1} + \cdots + n_e) Z_p$.◇

4.2.1.3 Remark

Let p be a prime number, K a perfect field, $char(K) = p$ and G an Abelian p-group. For all $n \in \mathbb{N}$ the statement $(rad(KG)^*)^{p^n} = rad(K(G^{p^n}))$ is true. In particular, G and $rad(KG)^*$ are of the same exponent.

Proof. The remark is a consequence of proposition 2.4.5 and corollary 2.4.4.◇

We are ready to determine the invariants for an Abelian group algebra. Let $soc(G)$ be the socle of an Abelian p-group G which is the largest elementary-p-Abelian subgroup of G. Recall that the rank of an Abelian p-group G is the dimension of the socle of G as $GF(p)$-space.

4.2.1.4 Theorem (invariants of Abelian unit groups)

Let $e, k \in \mathbb{N}$, p a prime number, K a field of order p^k and G an Abelian p-group of exponent p^e. For all $i \in \underline{e}$ let

The invariants of the center 105

$$s_i := k(\mid G^{p^{i-1}} \mid -2 \mid G^{p^i} \mid + \mid G^{p^{i+1}} \mid).$$

The group $rad(KG)^*$ is $\mathcal{G}$-isomorphic to $s_1 Z_p \times \cdots \times s_e Z_{p^e}$. In particular, $k(\mid G \mid - \mid G^p \mid)$ is the rank of $rad(KG)^*$ and $soc(rad(KG)^*) \cong_{\mathcal{G}} (k(\mid G \mid - \mid G^p \mid))Z_p$ is valid.

Proof. Because of remark 4.2.1.3 we derive $exp(G) = exp(rad(KG)^*)$, and for all $i \in \underline{e-1} \cup \{0\}$ the equation

(1) $\mid (rad(KG)^*)^{p^i}/(rad(KG)^*)^{p^{i+1}} \mid = \mid K \mid^{\mid G^{p^i} \mid - \mid G^{p^{i+1}} \mid}$

is true. For all $i \in \underline{e}$ let $s_i \in \mathbb{N}_0$ such that $rad(KG)^*$ is $\mathcal{G}$-isomorphic to $s_1 Z_p \times \cdots \times s_e Z_{p^e}$. We use (1) and proposition 4.2.1.2 to deduce

(2) $\forall i \in \underline{e} : s_i + \cdots + s_e = k(\mid G^{p^{i-1}} \mid - \mid G^{p^i} \mid).$

This system of linear equation can be solved straightforward. Thus, the first part of the theorem is valid. The additions are valid by using (2) in the case $i = 1$. ◇

4.2.1.5 Examples

Let p be a prime number, $e, n \in \mathbb{N}$, G an Abelian p-group, $exp(G) = p^e$, K a field, $\mid K \mid = p^k$ and $s_1, \ldots, s_e$ as presented within theorem 4.2.1.4.

(1) If $G = Z_{p^e}$ is valid, then for all $i \in \underline{e-1}$ the identities $s_i = kp^{e-i-1}(p-1)^2$ and $s_e = k(p-1)$ are valid. In particular, for $p = 2$ we derive $(s_e, s_{e-1}, \ldots, s_1) = k(1, 1, 2, 4, \ldots, 2^{e-2})$.

(2) If $G = nZ_{p^e}$ is valid, then for all $i \in \underline{e-1}$ the identities $s_i = kp^{(e-i-1)n}(p^n - 1)^2$ and $s_e = k(p^n - 1)$ are true.

(3) If $G = Z_p \times Z_{p^2} \times \cdots \times Z_{p^e}$ is true, then for all $i \in \underline{e-1}$ the identities $s_i = kp^{(e-i-1)(e-i)0.5}(p^{2(e-i)+1} - 2p^{e-i} + 1)$ and $s_e = k(p-1)$ are valid. In particular, for $p = 3$, $k = 1$ and $e = 4$ we derive $(s_1, s_2, s_3, s_4) = (57618, 678, 22, 2)$.

(4) If $G = nZ_p \times \cdots \times nZ_{p^e}$, then for all $i \in \underline{e-1}$ the statements $s_i = kp^{(e-i-1)(e-i)0.5}(p^{n(2(e-i)+1)} - 2p^{n(e-i)} + 1)$ and $s_e = k(p^n - 1)$ are true. ◇

Within these examples we derive that the invariants possess a distinctive monotone behavior. We will investigate this topic in more details within the next subsections. Moreover, we will prove that G is a direct factor of $1_G + rad(KG)$. In particular, for an arbitrary p-group G we derive that $Z(G)$ is a direct factor within $Z(1_G + rad(KG))$.

4.2.2 Complements

4.2.2.1 Proposition

Let p be a prime number, K a field, $char(K) = p$, G a p-group and (A, B) a direct decomposition of G. $(rad(KB)KG, rad(KA))$ is a semi-direct decomposition of the K-algebra $rad(KG)$.

Proof. By using definition and remark 1.1.14 we derive $ker\, p_B = KG\, rad(KB) = rad(KB)\, KG$. For all $a_1, a_2 \in A$ the statement $a_1 B = a_2 B$ is true if and only if $a_1 = a_2$ is valid. Thus, $rad(KA) \cap ker\, p_B = \{0_{KG}\}$ is proven. We finish the proof by using a dimension argument.⋄

4.2.2.2 Definition (decomposable without gap)

Let p a prime number, G an Abelian p-group, $e \in \mathbb{N}_0$ and $s_i \in \mathbb{N}$ for all $i \in \underline{e}$ such that G is $\mathcal{G}$-isomorphic to $s_1 Z_p \times \cdots \times s_e Z_{p^e}$. We call G decomposable without gap if for all $i \in \underline{e}$ the number s_i is not equal to zero.⋄

4.2.2.3 Theorem (decomposable without gap)

Let p be a prime number, G an Abelian p-group, K a finite field and $char(K) = p$. $rad(KG)^*$ is decomposable without gap.

Proof. Let $g \in G$ such that $o(g) = max\{o(x) \mid x \in G\}$. It is well-known (principle of maximal factors in Abelian groups) that $M := \langle g \rangle_{\mathcal{G}}$ possesses a complement in G. We use proposition 4.2.2.1 and corollary 1.1.8 to deduce that $rad(KM)^*$ is a direct factor of $rad(KG)^*$. The proof is finished by using part (1) of the examples 4.2.1.5 because the direct factor is already decomposable without gap and the exponent is reached by using a maximal factor (see remark 4.2.1.3).⋄

4.2.2.4 Remark

Let G be a finite group and U, V subgroups of G. Because of $\mid UV \mid \cdot \mid U \cap V \mid = \mid U \mid \cdot \mid V \mid$ the identity $\mid G/_r(U \cap V) \mid \leq \mid G/_r U \mid \cdot \mid G/_r V \mid$ is valid.⋄

4.2.2.5 Theorem (complements, Johnson, 1978)

Let p be a prime number, G an Abelian p-group, K a finite field and $char(K) = p$. G is a direct factor of $1_G + rad(KG)$.

Proof. (see [33]) We prove the theorem by an induction argument based on the order of G. Because of theorem 4.2.1.4 and corollary 2.4.4 the groups G and $1_G + rad(KG)$ are of the same exponent. If G is cyclic, then G is a

The invariants of the center 107

maximal factor of $1_G + rad(KG)$ and the theorem is proven.
Let G be non-cyclic. Thus, a non-trivial direct decomposition (A, B) of G exists. By using proposition 4.2.2.1 we deduce

(1) $rad(KG) = (rad(KA)KG) \oplus_K rad(KB) = (rad(KB)KG) \oplus_K rad(KA)$.

Corollary 1.1.8 and induction is used to derive that A resp. B is a direct factor of $1_G + rad(KA)$ resp. of $1_G + rad(KB)$. Let N_A resp. N_B a complement of A resp. B in $1_G + rad(KA)$ resp. in $1_G + rad(KB)$. We define

(2) $N := ((1_G + rad(KA)KG)N_B) \cap ((1_G + rad(KB)KG)N_A)$

and prove that N is a complement of G in $1_G + rad(KG)$. We use remark 4.2.2.4 and (1) to deduce that the index of N in $1_G + rad(KG)$ is not greater than $\mid A \mid \cdot \mid B \mid = \mid G \mid$. In addition, we use (1) to derive the equations $((1_G + rad(KA)KG)N_B) \cap G = A$ and $((1_G + rad(KB)KG)N_A) \cap G = B$. Hence, the proof is finished.⋄

4.2.2.6 Corollary (direct factor of the center)

Let p be a prime number, G a p-group, K a finite field and $char(K) = p$. $Z(G)$ is a direct factor of $Z(1_G + rad(KG))$.

Proof. The corollary is a consequence of theorem 4.2.2.5 and part (iv) of corollary 4.1.5.⋄

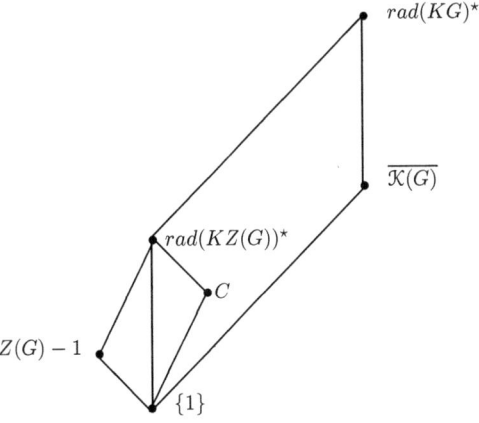

4.2.2.7 Corollary (cyclic center)

Let p be a prime number, G a non-trivial p-group, K a field and $char(K) = p$. $Z(rad(KG)^*)$ is cyclic if and only if $|G| = 2 = |K|$ is valid.

Proof. Let $Z(rad(KG)^*)$ be cyclic. Based on proposition 2.4.7 we derive that K is finite. By using corollary 4.2.2.6 we deduce that $Z(G) = Z(rad(KG))+1$ is valid. Thus, all conjugacy classes are of length 1 and KG is commutative. Hence, $1_G + rad(KG)$ is cyclic, too. Theorem 4.2.2.5 shows us that G is a direct factor of $1_G + rad(KG)$. Therefor, $G = 1_G + rad(KG)$ must be valid. This statement is – by using remark 1.1.18 – only valid for the mentioned values of G and K.
If $|G| = 2 = |K|$ is true, then $1_G + rad(KG) = G$ is $\mathcal{G}$-isomorphic to Z_2. ◊

4.2.3 Monotony

4.2.3.1 Definition (monotone decomposition)

Let p be a prime number, G an Abelian p-group, $e \in \mathbb{N}$ and $s_i \in \mathbb{N}$ for all $i \in \underline{e}$ such that G is $\mathcal{G}$-isomorphic to $s_1 Z_p \times \cdots \times s_e Z_{p^e}$. We call G monotone resp. strict monotone decomposable if and only if $s_1 \geq \cdots \geq s_e$ resp. $s_1 > \cdots > s_e$ is valid. ◊

4.2.3.2 Theorem (monotone decomposition)

Let p be a prime number, K a finite field, $char(K) = p$, G a group $\mathcal{G}$-isomorphic to $t_1 Z_p \times \cdots \times t_e Z_{p^e}$ and s_i ($i \in \underline{e}$) as in theorem 4.2.1.4.

(i) Except in the case $t_e = 1$, $t_{e-1} = 0$ for all $i \in \underline{e-1}$, the inequality $s_i \geq p \cdot s_{i+1}$ is valid.

(ii) $rad(KG)^*$ is strict monotone decomposable for $p \neq 2$.

(iii) Except in the case $t_e = 1$, $t_{e-1} = 0$ the group $rad(KG)^*$ is strict monotone decomposable for $p = 2$.

(iv) $rad(KG)^*$ is monotone decomposable for $p = 2$.

Proof. The proof is straightforward to be executed by using the invariants s_i, $i \in \underline{e}$ described in theorem 4.2.1.4. ◊

The invariants of the center

4.3 The invariants

4.3.1 The chain of Frattini subgroups

4.3.1.1 Lemma

Let p be a prime number, K a perfect field, $char(K) = p$ and G a p-group. For all $i \in \mathbb{N}$ the set $\{\overline{(g^{p^i})^G} \mid g \in G \setminus Z(G), C_G(g) = C_G(g^{p^i})\}$ is a K-basis of the K-space $(\overline{\mathcal{K}(G)}^*)^{p^i}$.

Proof. The set $\overline{\mathcal{K}(G)}$ is K-generated by the conjugacy class sums of G and is central in $rad(KG)$. For all $C \in \mathcal{K}(G)$ let $k_C \in K$. Because of $char(K) = p$ we derive:

(1) $(\sum_{C \in \mathcal{K}(G)} k_C \overline{C})^{p^i} = \sum_{C \in \mathcal{K}(G)} (k_C)^{p^i} (\overline{C})^{p^i}$.

We use (1), the perfectness of K and corollary 2.4.4 to deduce that $(\overline{\mathcal{K}(G)}^*)^{p^i}$ is a K-space which is K-generated by the set $\{\overline{(g^G)}^{p^i} \mid g \in G \setminus Z(G)\}$. Finally, we use theorem 2.4.8 – which is based on the concept of end-commutable orderings – to finish the proof.⋄

4.3.1.2 Definition (special dimension)

Let p be a prime number, K a field, $char(K) = p$ and G a p-group. For all $i \in \mathbb{N}_0$ we define

$$\overline{k(G)}_{p^i} := |\{\overline{(g^{p^i})^G} \mid g \in G \setminus Z(G), C_G(g) = C_G(g^{p^i})\}|.\diamond$$

Based on lemma 4.3.1.1 we derive that $\overline{k(G)}_{p^i}$ is the K-dimension of $(\overline{\mathcal{K}(G)}^*)^{p^i}$.

4.3.1.3 Theorem (invariants of the center, I)

Let p be a prime number, K a finite field of order p^k, G a p-group and $s_1, \ldots, s_e \in \mathbb{N}_0$ such that $\overline{\mathcal{K}(G)}^*$ is $\mathcal{G}$-isomorphic to $s_1 Z_p \times \cdots \times s_e Z_{p^e}$. For all $i \in \underline{e}$ the identity

$$s_i = k(\overline{k(G)}_{p^{i-1}} - 2 \cdot \overline{k(G)}_{p^i} + \overline{k(G)}_{p^{i+1}})$$

is valid.

Proof. For all $i \in \underline{e}$ we deduce – based on lemma 4.3.1.1 – the statement

(1) $|(\overline{\mathcal{K}(G)}^*)^{p^i}| = p^{k \cdot \overline{k(G)}_{p^i}}$.

Proposition 4.2.1.2 lets us deduce that for all $i \in \underline{e-1} \cup \{0\}$ the identity

(2) $(\overline{\mathcal{K}(G)}^*)^{p^i}/(\overline{\mathcal{K}(G)}^*)^{p^{i+1}} \cong_{\mathfrak{g}} (s_{i+1} + \cdots + s_e)Z_p$

is true. We use (1) and (2) to derive for all $i \in \underline{e-1} \cup \{0\}$ the statement

(3) $k(\overline{k(G)}_{p^i} - \overline{k(G)}_{p^{i+1}}) = s_{i+1} + \cdots + s_e$

This system of linear equations can be solved as desired.◇

The order of K is increasing the invariants in a smooth way. The invariants different from zero are determined purely by the group G. It is not known whether all invariants are different from zero or monotone as proven for the Abelian case.

4.3.1.4 Example

Let K be a field of order 2^k and G the dihedral group of order 16. Elements $h, a \in G$ exist such that $o(h) = 8$, $o(a) = 2$, $G = \langle h, a \rangle_{\mathfrak{g}}$ and $h^a = h^{-1}$ are valid. The non-central conjugacy classes of G are a^G, $(ha)^G$, h^G, $(h^3)^G$ and $(h^2)^G$.

a, ha are involutions, and thus based on theorem 2.4.8 we deduce $(\overline{a^G})^2 = ((\overline{ha})^G)^2 = 0_{KG}$. $\langle h^4 \rangle_{\mathfrak{g}} = Z(G)$ and $\langle h \rangle_{\mathfrak{g}} = C_G(h) = C_G(h^3) = C_G(h^2) = C_G(h^6)$ are valid. Therefor, theorem 2.4.8 lets us derive that $(\overline{h^G})^2 = \overline{(h^2)^G} = ((\overline{h^3})^G)^2$ is true.
We determine $\overline{k(G)}_{2^0} = 5$, $\overline{k(G)}_{2^1} = 1$ and $\overline{k(G)}_{2^2} = 0$, and by using theorem 4.3.1.3 we conclude $\overline{\mathcal{K}(G)}^* \cong_{\mathfrak{g}} (3k)Z_2 \times (1k)Z_4$.◇

4.3.2 The chain of socle subgroups

4.3.2.1 Definition and remark (n-th socle)

Let p be a prime number, $n \in \mathbb{N}_0$ and G an Abelian p-group. We define $soc_n(G) = \{g \mid g \in G, g^{p^n} = 1_G\}$ and call $soc_n(G)$ the n-th socle of G. For all $n \in \mathbb{N}$ the set $soc_n(G)$ is a subgroup of $soc_{n+1}(G)$.◇

4.3.2.2 Examples

Let p be a prime number, K a field, $char(K) = p$ and G a p-group.

(i) If U is a non-trivial subgroup of G, then the identity $\overline{U}^2 = \mid U \mid_K \overline{U} = 0_{KG}$ is valid. In particular, $\overline{G}$ and $\overline{Z(G)}$ are contained in the 1-st socle of $Z(rad(KG)^*)$ (see corollary 2.4.4).

(ii) Let p^n be the maximal length of all conjugacy classes of G. For all

$i \in \underline{n}$ let C_{p^i} the union of all conjugacy classes of length p^i of G. By using (i) we deduce $0_{KG} = \overline{G}^p = (\overline{Z(G)} + \sum_{i=1}^{n} \overline{C_{p^i}})^p = \sum_{i=1}^{n} \overline{C_{p^i}}^p$. Thus, the element $\bigcup_{i=1}^{n} \overline{C_{p^i}}$ is contained in the 1-st socle of $Z(rad(KG)^*)$ by using corollary 2.4.4. We use theorem 2.4.8 to deduce that for all $C \in \mathcal{K}(G)$ either the statement $\overline{C}^p = 0_{KG}$ or $\overline{C}^p = \overline{C^p}$ and $\mid C \mid = \mid C^p \mid$ are valid. Hence, for every $i \in \underline{n}$ the element $\overline{C_{p^i}}$ is contained in the 1-st socle of $Z(rad(KG)^*)$ (see corollary 2.4.4).

(iii) If $g \in G$ is of order p, then $\overline{g^G}$ is an element of order p based on theorem 2.4.8. In particular, for every involution g of G the element $\overline{g^G}$ is an involution, too.◊

Straightforward to prove is the following fact:

4.3.2.3 Proposition

Let p be a prime number, $s_1, \ldots, s_e \in \mathbb{N}$ and G a group $\mathcal{G}$-isomorphic to $s_1 \mathbb{Z}_p \times \cdots \times s_e \mathbb{Z}_{p^e}$. For all $i \in \underline{e}$ the identity $\mid soc_i(G)/soc_{i-1}(G) \mid = p^{s_i + \cdots + s_e}$ is valid.◊

4.3.2.4 Definition and remark (special equivalence relation)

Let K be a field, G a group and $n \in \mathbb{N}$. For all non-central conjugacy classes C, D of G let $\overline{C} \sim_n \overline{D}$ defined by $\overline{C}^n = \overline{D}^n$. $\sim_n$ is an equivalence relation on the set of all non-central conjugacy classes of G.◊

4.3.2.5 Lemma

Let p be a prime number, $n, r \in \mathbb{N}$, K a field, $char(K) = p$, G a p-group and $L_1, \ldots, L_r$ the equivalence classes of $\sim_{p^n}$. W.l.o.g. we assume $\overline{C}^{p^n} = 0_{KG}$ for all $C \in L_1$. The set $\langle L_1 \rangle_K \oplus_K \bigoplus_{i=2}^{r} {}_K Aug_{L_i}(\langle L_i \rangle_K)$ is the n-th socle of $\overline{\mathcal{K}(G)}^*$.

Proof. For all $i \in \underline{r}$ let $C_i \in L_i$. If $z \in \overline{\mathcal{K}(G)}$ is valid, then for every $i \in \underline{r}$ and $D_i \in L_i$ an element $k_{D_i} \in K$ exists such that

(1) $z = \sum_{i=1}^{r} \sum_{D_i \in L_i} k_{D_i} \overline{D_i}$

is true. Because of $char(K) = p$ and (1) the statement

(2) $z^{p^n} = \sum_{i=2}^{r} (\sum_{D_i \in L_i} k_{D_i})^{p^n} \overline{C_i}^{p^n}$

is valid. Based on (2) the identity $z^{p^n} = 0_{KG}$ is equivalent to the fact that for all $i \in \underline{r} \setminus \{1\}$ the statement $(\sum_{D_i \in L_i} k_{D_i})^{p^n} = 0_{KG}$ is valid. The Frobenius homomorphism is injective. Hence, the proof is finished by using corollary 2.4.4.⋄

4.3.2.6 Theorem (invariants of the center, II)

Let p be a prime number, K a finite field of order p^k, G a p-group, p^e the exponent of $\overline{\mathcal{K}(G)}^*$, for all $i \in \underline{e}$ the element l_i the number of equivalence classes of $\sim_{p^i}$ and $\overline{\mathcal{K}(G)}^*$ $\mathcal{G}$-isomorphic to $s_1 Z_p \times \cdots \times s_e Z_{p^e}$. $s_e = k(l_{e-1} - l_e)$, $s_i = k(l_{i-1} - 2l_i + l_{i+1})$ for all $2 \leq i \leq e-1$ and $s_1 = k(c(G) - |Z(G)| - 2l_1 + 1 + l_2)$ are valid.

Proof. For all $i \in \underline{e}$ the i-th socle of $\overline{\mathcal{K}(G)}^*$ is – based on lemma 4.3.2.5 – of order $p^{k(c(G) - |Z(G)| - (l_i - 1))}$. Proposition 4.3.2.3 let us derive the identities

$s_e = k(l_{e-1} - l_e)$,
$s_i + \cdots + s_e = k(l_{i-1} - l_i)$ for all $2 \leq i \leq e-1$ and
$s_1 + \cdots + s_e = k(c(G) - |Z(G)| - l_1 + 1)$.

This system of linear equations can be solved as desired.⋄

4.3.2.7 Remark

Let p be a prime number, K a field, $char(K) = p$, G a p-group, $n \in \mathbb{N}$ and $g, h \in G \setminus Z(G)$. Because of theorem 2.4.8 the statement $g^G \sim_{p^n} h^G$ is valid if and only if $(C_G(g) = C_G(g^{p^n})$ and $C_G(h) = C_G(h^{p^n}))$ or $(C_G(g) < C_G(g^{p^n})$ and $C_G(h) < C_G(h^{p^n}))$ are true.⋄

4.3.2.8 Example

As within example 4.3.1.4 let K be a field of order 2^k and G a dihedral group of order 16. Let h, a as in example 4.3.1.4. Remark 4.3.2.7 lets us deduce that $\{h^G, (h^3)^G\}$ and $\{a^G, (ha)^G, (h^2)^G\}$ are the equivalence classes of $\sim_{2^1}$. In addition, all non-central classes are equivalent with respect to $\sim_{2^2}$. We use the definitions within theorem 4.3.2.6 and deduce $l_1 = 2$ and $l_2 = 1$. Hence, again by using theorem 4.3.2.6 we conclude that $s_2 = k(2-1) = 1k$ and $s_1 = k(5 - 4 + 1 + 1) = 3k$ are valid.⋄

4.4 The class-graph

The graph used in this section is visualizing the determination of the exponent and the invariants of the center of $rad(KG)^*$ for a p-group G and a finite field K with $char(K) = p$.

4.4.1 Definition and remark (class graph)

Let p be a prime number, K a field, $char(K) = p$ and G a p-group. We define an oriented graph which we call the class-graph of G. Its set of vertices is $\{\overline{g^G} \mid g \in G \setminus Z(G)\} \cup \{0_{KG}\}$, and its set of edges is $\{(\overline{g^G}; \overline{(g^G)^p}) \mid g \in G \setminus Z(G)\}$. The class-graph can be determined by using theorem 2.4.8 in which the powering with p is described on the set of conjugacy class sums. Within the next example we demonstrate the calculation of the exponent and of the invariants of the center of the radical of KG with respect to $*$. The argumentation used in the example is true in general and can be applied to every example.◇

4.4.2 Example

Let K be a field of order 2^k and G the dihedral group of order 32. In this case $p = 2$ is valid. Let $h, a \in G$ such that $o(a) = 2$, $o(h) = 16$ and $h^a = h^{15}$ are valid. The non-central conjugacy classes of G are a^G, $(ha)^G$ and $(h^i)^G$ for all $i \in \mathbb{Z}$. The class-graph of G is – based on theorem 2.4.8 – of the following shape:

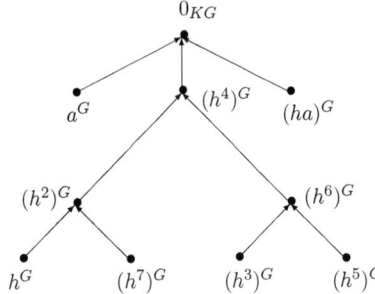

(i) The determination of the exponent:
Let $g \in G \setminus Z(G)$. There is exactly one path from $\overline{g^G}$ to the vertex 0_{KG}. If n is the length of this path, then $o(\overline{g^G}) = p^n = 2^n$ is valid. For example, $\overline{h^G}$ is of order 2^3, and this is the maximum of the set $\{o(\overline{g^G}) \mid g \in G \setminus Z(G)\}$. Hence, it is the exponent of $\overline{\mathcal{K}(G)}^*$ (see proposition 2.4.7 and example 3.1.2).

If m is the largest way from a vertex different from zero to the vertex 0, then the exponent of $\overline{\mathcal{K}(G)}^*$ is exactly $p^m = 2^m$.

(ii) **The determination of the invariants, I:**
In this case the invariants are determined by using the chain of Frattini subgroups. Let $e \in \mathbb{N}$, $2^e = p^e = max\{o(\overline{g^G}) \mid g \in G \setminus Z(G)\}$ and $n \leq e$. From every vertex $\neq 0_{KG}$ there is exactly one path of length n. The number of end-vertices different from 0_{KG} of all these pathes (for every vertex different from zero) is the number $\overline{k(G)}_{p^n}$. In this example we obtain $\overline{k(G)}_{2^0} = 9$, $\overline{k(G)}_{2^1} = 3$, $\overline{k(G)}_{2^2} = 1$ and $\overline{k(G)}_{2^3} = 0$. If we use the notations of theorem 4.3.1.3, then we deduce $s_1 = k(9 - 2 \cdot 3 + 1) = 4k$, $s_2 = k(3 - 2 \cdot 1 + 0) = k$ and $s_3 = k(1 - 2 \cdot 0 + 0) = k$.

(iii) **The determination of the invariants, II:**
In this case the invariants are determined by using the chain of socles. Let $e \in \mathbb{N}$ such that $p^e = max\{o(\overline{g^G}) \mid g \in G \setminus Z(G)\}$ and $n \leq e$ are valid. From every vertex $\neq 0_{KG}$ there is exactly one path of length n. Two vertices $\neq 0_{KG}$ are equivalent with respect to $\sim_{p^n}$ if the end-vertices of these pathes are identical. In our example there are $l_1 = 4$ classes for $\sim_{2^1}$ – which are $\{h^G, (h^7)^G\}$, $\{(h^3)^G, (h^5)^G\}$, $\{(h^2)^G, (h^6)^G\}$ and $\{a^G, (ha)^G, (h^4)^G\}$ –, $l_2 = 2$ classes for $\sim_{2^2}$ – which are $\{h^g, (h^3)^G, (h^5)^G, (h^7)^G\}$ and $\{a^g, (ha)^G, (h^2)^G, (h^4)^G, (h^6)^G\}$ – and $l_3 = 1$ class for $\sim_{2^3}$. We use the notations of theorem 4.3.2.6 and deduce $s_3 = k(2-1) = k$, $s_2 = k(4 - 2 \cdot 2 + 1) = k$ and $s_1 = k(9 - 2 \cdot 4 + 1 + 2) = 4k$.

(iv) We remark that – because the p-powering on conjugacy class sum is 0 or a class sum of the same length – the class-graph can be grouped by the class sums to conjugacy classes of the same length.◊

An easy application of theorems 4.3.1.3 and 4.3.2.6 is the following theorem:

4.4.3 Theorem (center and class graphs)

Let p be a prime number, K a field, $char(K) = p$ and G, H be two p-groups possessing isomorphic class-graphs. Then $(\overline{\mathcal{K}(G)}^*)$ and $(\overline{\mathcal{K}(H)}^*)$ are isomorphic.◊

4.5 Determination of invariants for special cases

4.5.1 The minimal case

4.5.1.1 Theorem (elementary-Abelian co-factor)

Let p be a prime number, K a finite field of order p^k and G a p-group such that for all $g \in G \backslash Z(G)$ the identity $C_G(g) < C_G(g^p)$ is valid. $Z(rad(KG)^*)$ is $\mathcal{G}$-isomorphic to $(rad(KZ(G))^*) \times (k(c(G) - |Z(G)|)Z_p)$.

Proof. Based on part (iii) of corollary 4.1.5 the group $Z(rad(KG)^*)$ is $\mathcal{G}$-isomorphic to $rad(KZ(G))^* \times \overline{\mathcal{K}(G)}^*$. The Abelian group $\overline{\mathcal{K}(G)}^*$ is of exponent p because of theorem 2.4.8. By definition this group is of order $p^{k(c(G)-|Z(G)|)}$, and this finishes the proof.⋄

4.5.1.2 Examples (regular p-groups, special p-groups)

Let p be a prime number, K a finite field of order p^k and G a p-group.

(i) If G^p is central in G, then G is usable (based on theorem 2.4.8) for theorem 4.5.1.1. For example, G^p is central if G is a special or minimal non-Abelian p-group (see proposition 3.3.4).

(ii) The regular p-groups are –based on theorem 2.4.8 and corollary 3.3.2 – fulfilling the prerequisites of theorem 4.5.1.1.

(iii) Let $Z(G)$ be elementary-Abelian and $|G'| = p$ be valid. Based on part (ii) of corollary 3.3.2 is the group $Z(rad(KG)^*)$ elementary-Abelian. We use a theorem within [36] to deduce that every non-central conjugacy class if of length p. Hence, exactly $\frac{|G|-|Z(G)|}{p}$ conjugacy classes of length p are existing and $c(G) = \frac{|G|+(p-1)|Z(G)|}{p}$ is valid. We use theorem 4.5.1.1 to conclude that $Z(rad(KG)^*)$ is $\mathcal{G}$-isomorphic to $(k^{\frac{|G|+(p-1)|Z(G)|-p}{p}})Z_p$.

(iv) Let $Z(G) = G' \cong_{\mathcal{G}} Z_p$ be valid. Based on part (iii) we deduce that $Z(rad(KG)^*)$ is $\mathcal{G}$-isomorphic to $(k(\frac{|G|}{p}+p-2))Z_p$. For example, the extra-special p-groups are applicable to part (iv).

(v) Let G be non-Abelian and of order p^3. Part (iv) lets us deduce that $Z(rad(KG)^*)$ is $\mathcal{G}$-isomorphic to $(k(p^2+p-2))Z_p$.⋄

4.5.2 The maximal case

Let p be a prime number, K a finite field of order p^k and G a non-Abelian p-group such that $exp(Z(rad(KG)^*)) = \frac{|G|}{p^2}$ is valid. Because of corollary 2.5.3

the value $\frac{|G|}{p^2}$ is the maximal possible value for $exp(Z(rad(KG)^*))$. Based on theorem 3.1.6 we deduce that $exp(Z(G)) = \frac{|G|}{p^2}$ is true or G possesses a cyclic maximal subgroup.◇

4.5.2.1 The case $exp(Z(G)) = \frac{|G|}{p^2}$

Let $n \in \mathbb{N}_{\geq 3}$ such that $|G| = p^n$ is valid. Based on $|Z(G)| = p^{n-2}$ we deduce that every non-central conjugacy class of G is of length p. Hence, their number is $\frac{|G|-|Z(G)|}{p} = p^{n-1} - p^{n-3}$. $G/Z(G)$ is not cyclic, and thus $G^p \subseteq Z(G)$ is valid. Based on theorem 2.4.8 we obtain that $\overline{\mathcal{K}(G)}^*$ is elementary-Abelian of order $p^{k(p^{n-1}-p^{n-3})}$. $Z(G) \cong_{\mathcal{G}} Z_{p^{n-2}}$ is true, and thus we can use theorem 4.2.1.4 and part (iii) of corollary 4.1.5 to determine the following decomposition of $Z(rad(KG)^*)$ into cyclic p-groups:

$(k(p-1))Z_{p^{n-2}}$
$\times (kp^{i-3}(p-1)^2)Z_{p^{n-i}}$ (for all $i \in \underline{n-3} \setminus \underline{2}$)
$\times (k(p^{n-4}(p-1)^2 + p^{n-1} - p^{n-3}))Z_p$.◇

In the next section we analyse the case that G possesses a cyclic maximal subgroup. Based on the statement within the pages 98 und 99 in [71] we know that G is a dihedral, a semi-dihedral, a quaternion group or a group of part (a) or (d) of theorem 5.3.2 in [71]. The last two groups are possessing a large center and are already considered within the case 4.5.2.1.

4.5.2.2 Dihedral, semi-dihedral and quaternion groups

We begin the analysis with the dihedral and quaternion groups.
Let $p = 2$, $n \in \mathbb{N}_{\geq 3}$ and $h, a \in G$ such that $G = \langle h, a \rangle_{\mathcal{G}}$, $o(h) = 2^{n-1}$, $o(a) = 2$ and $h^a = h^{-1}$. The conjugacy classes of G are $\{1_G\}$, $\{h^{2^{n-2}}\}$, a^G, $(ha)^G$ and $\{h^r, h^{-r}\}$ for $r \in \underline{2^{n-2}-1}$. There are $2^{n-2} + 1$ non-central conjugacy classes. Based on theorem 4.2.1.4 we deduce $rad(KZ(G))^* \cong_{\mathcal{G}} kZ_2$. Now we are determining the group $\overline{\mathcal{K}(G)}^*$. Based on the examples 3.1.2 we know that the exponent of the group is exactly 2^{n-2}. Let $i \in \underline{n-2}$. If G is a dihedral group, then a, ha are involutions, in the other case a, ha are elements of order 4 such that their squares are central in G. Therefor, in both cases theorem 2.4.8 shows us the identity $(\overline{a^G})^{2^i} = 0_{KG} = (\overline{(ha)^G})^{2^i}$. Let $r \in \underline{2^{n-2}-1}$. The elements h^r and h^{-r} are commuting with each other, and thus $(h^r + h^{-r})^{2^i} = h^{2^i r} + h^{-2^i r}$ is true. Within the normal subgroup $\langle h^{2^i} \rangle_{\mathcal{G}}$ exactly $2^{n-i-2} - 1$ non-central conjugacy classes of G are contained. We deduce $\overline{k(G)}_{2^i} = 2^{n-i-2} - 1$ and $\overline{k(G)}_{2^0} = 2^{n-2} + 1$. For all $i \in \underline{n-2}$ let s_i the number of factors $\mathcal{G}$-isomorphic to Z_{2^i} in a direct decomposition of $\overline{\mathcal{K}(G)}^*$ into cyclic 2-groups. We use theorem 4.3.1.3 to conclude $(s_{n-2}, s_{n-3}, s_{n-4}, s_{n-5}, \ldots, s_2, s_1) = k(1, 1, 2^1, 2^2, \ldots, 2^{n-5}, 2^{n-4} + 2)$. For

The invariants of the center   117

all $i \in \underline{n-2}$ the element t_i be the number of factors $\mathcal{G}$-isomorphic to Z_{2^i} in a direct decomposition of $Z(rad(KG)^*)$ into cyclic 2-groups. Because of part (iii) of corollary 4.1.5 we deduce $(t_{n-2}, t_{n-3}, t_{n-4}, t_{n-5}, \ldots, t_2, t_1) = k(1, 1, 2^1, 2^2, \ldots, 2^{n-5}, 2^{n-4} + 3)$.

Now we are focussing on the semi-dihedral groups.
Let $h, a \in G$ such that $G = \langle h, a \rangle_{\mathcal{G}}$, $o(h) = 2^{n-1}$, $o(a) = 2$ and $h^a = h^{-1} h^{2^{n-2}}$ are true. Let $z := h^{2^{n-2}}$. $Z(G) = \langle z \rangle_{\mathcal{G}}$ is valid, and for all $r \in \mathbb{N}$ the identity $(h^r)^a = h^{-1} z^r$ is true. The conjugacy classes of G are $\{1_G\}$, $\{z\}$, a^G, $(ha)^G$ and $\{h^r, h^{-r} z^r\}$ for all $r \in \underline{2^{n-2} - 1}$. Because of $z^2 = 1_G$ the center of G is $\mathcal{G}$-isomorphic to Z_2. Let $i \in \underline{n-2}$. We use $a^2 = 1_G$, $(ha)^2 = z \in Z(G)$ and theorem 2.4.8 to obtain $\overline{(a^G)}^{2^i} = 0_{KG} = \overline{((ha)^G)}^{2^i}$. If $r \in \underline{2^{n-2} - 1}$, then – because of $z^2 = 1_G$ – the identity $(h^r + h^{-r} z^{-r})^{2^i} = h^{2^i r} + h^{-2^i r}$ is valid. We conclude that the invariants of $Z(rad(KG)^*)$ are the same as already presented for the dihedral and quaternion groups.◇

4.5.2.3 Isomorphism and non-isomorphism

The previous example shows us also the following remarkable observation: Two 2-groups G and H and a finite field K exist such that

(i) G and H are not isomorphic,

(ii) $Z(G)$ and $Z(H)$ are isomorphic and

(iii) $Z(1 + rad(KG))$ and $Z(1 + rad(KH))$ are isomorphic.

Deskins has proven in [22] that for finite Abelian p-groups G, H and for a finite field K of characteristic p the group algebras KG and KH – or the equivalent statement that the unit groups $E(KG)$ and $E(KH)$ are isomorphic – if and only if the groups G and H are isomorphic. A possible extension to non-Abelian groups would be that the centers $Z(G)$ and $Z(H)$ are isomorphic if and only if $Z(1 + rad(KG))$ and $Z(1 + rad(KH))$ are isomorphic. But this is not true, and for this fact we provide two examples now.

Within the first example take an arbitrary non-Abelian extra-special p-group G. We use example 4.5.1.2 to prove that $Z(rad(KG)^*)$ is $\mathcal{G}$-isomorphic to $(k(\frac{|G|}{p} + p - 2))Z_p$. If we take $H := Z(G)$, then we use example 4.2.1.5 to deduce that $Z(rad(KH)^*)$ is $\mathcal{G}$-isomorphic to $(k(p-1))Z_p$. Thus, the centers of G and H are isomorphic but $Z(rad(KG)^*)$ and $Z(rad(KH)^*)$ are not isomorphic.

For the second example we use [79]: two 3-groups G, H of exponent 3 with the same number of conjugacy classes but non-isomorphic centers exist. Indeed, the centers are isomorphic to $Z_3 \times Z_3$ and $Z_3 \times Z_3 \times Z_3$. The corresponding centers of $Z(rad(KG)^*)$ and $Z(rad(KH)^*)$ are both elementary-Abelian of the same size because of example 4.5.1.2 and corollary 4.1.5.

For understanding the structure of $Z(rad(KG)^*)$ it would be very helpful if we could prove that this group and/or the normal subgroup related to all non-central conjugacy class sums are presentable as an unit group of a group algebra based on an Abelian p-group. This is also not true, and we provide three examples for this topic.

Again, we focus on an arbitrary non-Abelian extra-special p-group G. We use example 4.5.1.2 to prove that $Z(rad(KG)^*)$ is $\mathcal{G}$-isomorphic to $(k(\frac{|G|}{p} + p - 2))Z_p$. If an Abelian p-group A would exists such that $Z(rad(KG)^*)$ and $1 + rad(KA)$ are isomorphic, then the order of A would be a p-power and identical to $\frac{|G|}{p} + p - 1$ which is a contradiction.

The second example is based on a p-group P of order p^3 for $p \neq 2$. Let $G = P \times P$. The center of G is of order p^2. P possesses the class number $c(P) = p^2 + p - 1$. Thus, the class number of G is $c(P)^2 = p^4 + 2p^3 - p^2 - 2p^1$. We deduce that $c(G) - \mid Z(G) \mid$ is exactly $p^4 + 2p^3 - 2p^2 - 2p + 1$. If the normal subgroup in the center of $1 + rad(KG)$ related to the class sums would be isomorphic to $1 + rad(KA)$ for an Abelian p-group A, then the equality $\mid A \mid = c(G) - \mid Z(G) \mid + 1 = p^4 + 2p^3 - 2p^2 - 2p + 2$ would be valid. The right hand side is not divisible by $p \neq 2$.

In the case $p = 2$ we use the group D_{2^n} for an element $4 \leq n \in \mathbb{N}$. Its class number is $c(D_{2^n}) = 2^{n-2} + 3$. If the normal subgroup in the center of $1 + rad(KG)$ related to the class sums would be isomorphic to $1 + rad(KA)$ for an Abelian p-group A, then the equality $\mid A \mid = 2^{n-2} + 2$ would be valid. The value on the right hand side is no power of 2.⋄

4.6 Isoclinism

In this section we investigate the exponent of the center within unit groups of modular group algebras under isoclinism. The results are used to describe the structure of the center for special types of groups like semi-extra-special, Camina and VZ-groups.

4.6.1 Isoclinic groups

The concept of isoclinic groups was introduced by P. Hall in 1940 (see [25]) for classifying p-groups. Within this article Hall has proven also some results related to stem groups and to subgroups which are supplements of the center.

4.6.2 Definition and remark (Isoclinism)

For a group G we define the commutator map $[,]_G : G/Z(G) \times G/Z(G) \longrightarrow G'$, $(Z(G)a; Z(G)b) \mapsto [a, b]$. Two groups G, H are called isoclinic if two isomorphism $\alpha : G/Z(G) \longrightarrow H/Z(H)$ and $\beta : G' \longrightarrow H'$ exist such that $(\alpha \times \alpha)[,]_H = [,]_G \beta$ is valid.

$$\begin{array}{ccc} G/Z(G) \times G/Z(G) & \xrightarrow{[,]_G} & G' \\ \alpha \times \alpha \downarrow & & \beta \downarrow \\ H/Z(H) \times H/Z(H) & \xrightarrow{[,]_H} & H' \end{array}$$

Isoclinism is an equivalence relation among isomorphism types of finite groups. Groups of minimal order in an isoclinism class are called stem groups and are characterized by the property that their center is contained in their derived subgroup. In addition, Hall has proven that every subgroup H – such that $HZ(G) = G$ is valid – is isoclinic to G.⋄

4.6.3 Theorem (exponents and isoclinic groups)

Let p be a prime number, K a field, $char(K) = p$ and G, H be two isoclinic p-groups. The groups $\overline{\mathcal{K}(G)}^*$ and $\overline{\mathcal{K}(H)}^*$ are of the same exponent.

Add-on: If $g \in G \setminus Z(G)$ and $h \in H$ such that $hZ(H) = (gZ(G))\alpha$, then $h \in H \setminus Z(H)$ and $o(\overline{g^G}) = o(\overline{h^H})$ are valid.⋄

Proof. Let $g \in G \setminus Z(G)$ and $h \in H$ such that $hZ(H) = (gZ(G))\alpha$. The function β is an isomorphism, and thus $h \in H \setminus Z(H)$ is valid. Let p^n be the order of the class sum of g^G. We use theorem 2.4.8 to determine the order of the class sum of h^H. At first we show that the order of h^H is not greater than the one of g^G. By a symmetry argument – because H and G are isoclinic based on α^{-1} and β^{-1} and $(hZ(H))\alpha^{-1} = gZ(G)$ – we deduce that both orders are identical. Therefor, the add-on and the theorem are proven.

Let $y \in C_G(g^{p^n}) \setminus C_G(g)$ and $z \in H$ such that $(yZ(G))\alpha = zZ(H)$. α is a homomorphisms, and thus $(g^{p^n} Z(G))\alpha = h^{p^n} Z(H)$ is true. We use the isoclinism rules and deduce: $1 = 1\beta = [y, g^{p^n}] = [yZ(G), g^{p^n} Z(G)]_G \beta = [zZ(H), h^{p^n} Z(H)]_H = [z, h^{p^n}]$. We have proven that $z \in C_H(h^{p^n})$ is valid. In addition, $1 \neq [y, g] = [yZ(G), gZ(G)]_G \beta = [zZ(H), hZ(H)]_H = [z, h]$ is true, and thus $z \notin C_H(h)$ is proven. By using theorem 2.4.8 we have deduced $o(\overline{g^G}) \geq o(\overline{h^H})$.⋄

4.6.4 Corollary

Let p be a prime number, K a field, $char(K) = p$ and G a non-Abelian p-group. The exponent of $\overline{\mathcal{K}(G)}^*$ is identical to the exponent of $\overline{\mathcal{K}(H)}^*$ for the following groups H:

(i) H is a stem group in the isoclinism class of G (H is a group such that $Z(H) \leq H'$ is valid.).

(ii) H is a subgroup of G such that $HZ(G) = G$ is valid.

(iii) H is a subgroup such that $Z(H) \leq \Phi(H)$ is valid. ⋄

Proof. By using theorem 4.6.3 and definition and remark 4.6.2 we obtain parts (i) and (ii) directly. For part (iii) we use part (ii) several times to derive a subgroup H such that H contains no proper subgroup V such that $VZ(H) = H$ is valid. In other words, for every subgroup V of H such that $VZ(H) = H$ is valid we obtain $V = H$. Let E be a generating set of H containing $Z(H)$. Then $H = \langle E \rangle_\mathcal{G} = \langle E \setminus Z(H) \rangle_\mathcal{G} Z(H)$ is valid. Therefor, $\langle E \setminus Z(H) \rangle_\mathcal{G} = H$ must be true. We conclude that $Z(H)$ can be omitted from every generating set of H. ⋄

4.6.5 Examples (semi-extra-special, ultra-special, Camina, generalized Camina and VZ-groups)

(i) Berkovich has proven within [7] that the dihedral, quasi-dihedral, and quaternion groups of order 2^n are isoclinic for $n \geq 3$. Thus, we get a new proof of the fact that the exponents of the centers of the radicals of the corresponding group algebras based on $\star$ are identical (see theorem 3.1.6 and examples 3.1.2).

(ii) Let p be a prime number, K a field of characteristic p, $n \in \mathbb{N}$, $G := Z_{p^n}$ and $H := Z_{p^n} \wr Z_{p^n}$. Based on corollary 3.5.14 and remark 4.2.1.3 the exponents of the centers of $rad(KG)^*$ and $rad(KH)^*$ are identical. G is an Abelian group and every group isoclinic to G is Abelian, too. Therefore, G and H are not isoclinic.

(iii) **Semi-extra-special and ultra-special groups**:
We use several facts about semi-extra-special groups mentioned in [45] and [46]. We say G is a semi-extra-special p-group if G is a p-group with the property that for every maximal subgroup N of $Z(G)$ the quotient G/N is an extra-special group. It is known that extra-special p-groups are semi-extra-special and that semi-extra-special p-groups are special. Let G be a semi-extra-special group. From corollary 3.3.2 we know that $exp(Z(rad(KG)^*)) = p$ is true. In addition, for every element $g \in G \setminus Z(G)$ the conjugacy class g^G is exactly $gZ(G)$. Thus, exactly $\frac{|G|}{|Z(G)|} - 1$ non-central conjugacy classes exist.

The invariants of the center 121

Based on theorem 4.1.4 we know that the center of the radical is of dimension $\frac{|G|}{|Z(G)|} + |Z(G)| - 2$. If K is a field with p^k elements, then $Z(rad(KG))^*$ is elementary-p-Abelian and isomorphic to $k(\frac{|G|}{|Z(G)|} + |Z(G)| - 2)$ copies of Z_p. Using the Universal Coefficients Theorem, one can prove that if p is an odd prime, then every semi-extra-special p-group is isoclinic to a unique (up to isomorphism) semi-extra-special p-group of exponent p. Ultra-special p-groups are semi-extra-special p-groups such that $(|Z(G)|)^3 = |G|$ is valid. In this case $\frac{|G|}{|Z(G)|} + |Z(G)| - 2 = (|Z(G)|)^2 + |Z(G)| - 2$ is true.

(iv): **VZ-groups**:
We use several facts about VZ-groups mentioned in [45] and [46]. We say that a group G is a VZ-group if every nonlinear irreducible character vanishes off of the center of G: $\xi(g) = 0$ for every nonlinear irreducible character ξ and all elements $g \in G \setminus Z(G)$. VZ-groups are characterized by the property that they are isoclinic to a semi-extra-special p-group and that for every element $x \in G \setminus Z(G)$ the conjugacy class of x is exactly xG'. Now we can use part (iii) and theorem 4.6.3 to deduce that for a VZ-p-group G the direct factor $\overline{(\mathcal{K}(G)^*)}$ is elementary-p-Abelian. If K is a finite field with p^k elements, then an analogue argumentation as done within (iii) shows us that $\overline{(\mathcal{K}(G)^*)}$ is isomorphic to $k(\frac{|G|-|Z(G)|}{|G'|})$ copies of Z_p.

(v): Within part (iii) and (iv) we could also use the following rule: Let N be a subgroup of G. N is normal in G if and only if $\overline{N}$ is central in KG. In addition, $(\overline{N})^2 = |N| \cdot \overline{N}$ is valid. As a consequence the order of $\overline{N}$ in $Z(rad(KG))^*$ is exactly p. For every element $g \in G \setminus Z(G)$ such that $g^G = gN$ is valid the order of $\overline{g^G}$ is exactly p, too, because $(\overline{g^G})^p = (\overline{gN})^p = g^p(\overline{N})^p = 0$ is true.

(vi): **(generalized) Camina groups**:
Following [45], [46], [47] and [48] a group G is called a Camina group if for every element $x \in G \setminus G'$ the conjugacy class of x is exactly xG'. It is stated there that a Camina 2-group has nilpotence class 2 and – when p is an odd prime – that a Camina p-group has nilpotence class 2 or 3. A Camina p-group of class 3 has the property that $G'/Z(G)$ is elementary-p-Abelian. We want to prove that the direct factor $\overline{\mathcal{K}(G)}^*$ is elementary-p-Abelian. Let G be of class 2. Then, $G' \leq Z(G)$ is valid. We deduce that for every $g \in G \setminus Z(G)$ the class g^G is exactly gG'. By using part (v) we get $(\overline{g^G})^p = 0$. Hence, $\overline{\mathcal{K}(G)}^*$ is isomorphic to direct product of $k(\frac{|G|-|Z(G)|}{|G'|})$ copies of Z_p if K is a field of order p^k. Now let the class of G be exactly 3. The class sums based on elements $g \in G \setminus G'$ are of order p based on the same argumentation. For the class sums based on elements of $G' \setminus Z(G)$ we use that $G'/Z(G)$ is elementary-p-Abelian. They are also of order p based on theorem 2.4.8.

The number of non-central class sums can be determined using results for generalized Camina groups. For all $g \in G' \setminus Z(G)$ the class of g^G is exactly $gZ(G)$. Now we can calculate the number of non-central conjugacy classes as $\frac{|G|-|G'|}{|G'|} + \frac{|G'|-|Z(G)|}{|Z(G)|}$. Thus, if K is a field of p^k elements, then the direct factor related to the class sums is isomorphic to $k(\frac{|G|}{|G'|} + \frac{|G'|}{|Z(G)|} - 2)$ copies of Z_p.

Following [46] a group G is a generalized Camina group if for every $x \in G \setminus (Z(G)G')$ the conjugacy class of x is exactly xG'. It is stated that generalized Camina groups are isoclinic to Camina groups. Thus, we can use the previous part and theorem 4.6.3 to deduce that the direct factor $\overline{\mathcal{K}(G)}^*$ is again elementary-p-Abelian. We have to count the non-central conjugacy classes. In the case of nilpotency class 2 – isoclinism preserves the nilpotency class – the group is a VZ-group. Thus, $\overline{\mathcal{K}(G)}^*$ is isomorphic to $k(\frac{|G|-|Z(G)|}{|G'|})$ copies of Z_p by using part (iv). The case of nilpotency class 3 is more complex. First, we get $\frac{|G|-|G'|Z(G)|}{|G'|}$ classes for the set $G \setminus (G'Z(G))$. It is known that for every element $g \in (G'Z(G)) \setminus Z(G)$ the conjugacy class g^G is exactly $g[G',G]$. Therefore we get $\frac{|G'Z(G)|-|Z(G)|}{|[G',G]|}$ classes for the set $(G'Z(G)) \setminus Z(G)$. By combining both numbers we deduce that $\overline{\mathcal{K}(G)}^*$ is isomorphic to $k(\frac{|G|}{|G'|} - \frac{|Z(G)|}{|G'\cap Z(G)|} + \frac{|G'||Z(G)|}{|G'\cap Z(G)||[G',G]|} - \frac{|Z(G)|}{|[G',G]|})$ copies of Z_p.◊

4.7 Open topics and exercises

Open-ended questions 4 *(i) What is the structure of the center of the radical of KG with respect to $*$ for wreath products, general group extensions, central products, direct products, subdirect products, groups of maximal class, powerful (embedded) groups etc.?*

(ii) Are the invariants of the center of the radical of KG with respect to $*$ without gap, monotone or strict monotone?

(iii) Is the set of conjugacy class sums for conjugacy classes of size not smaller than a lower bound a substructure of the center of KG?

(iv) Determine the structure of the center of the group of units of a modular group algebra for a nilpotent group.

(v) Determine the structure of the center of the unitary subgroup (see exercise 211).

(vi) Is the opposite implication of theorem 4.4.3 also valid?

(vii) Find relations of the invariants of the center of the radical for two isoclinic p-groups.

Excercise 126 Analyze remark 3.3.6 in more details for the structure of the center.

Excercise 127 Find two non-isoclinic groups with isomorphic centers of the corresponding radicals. (Tip: [79])

Excercise 128 Use [27] to prove that for a finite p-group G possessing only two conjugacy class length the identity $exp(G/Z(G)) = p$ is valid. What are the consequences for the structure of $Z(rad(KG))^\star$ for a finite field of characteristic p?

Excercise 129 Use the article [28] for collecting some results concerning p-groups possessing only two conjugacy class length which are $\{1, p\}$ or $\{1, p^2\}$. Describe the structure of $Z(rad(KG))^\star$ for a finite field of characteristic p. What is the influence of isoclinism within this exercise? (Tip: use exercise 128)

Excercise 130 Let K be a finite field of characteristic 2, $G := D_8$ and $H := G \times Z_2$. Prove that G, H are isoclinic. Determine the structure and the exponent of $Z(rad(KG))^\star$ and $Z(rad(KH))^\star$. Are both centers isomorphic?

Excercise 131 Determine the exponent of the center of the radicals of group algebras with respect to $\star$ for all p-groups mentioned in theorems A, B and C of [29].

Excercise 132 Prove definition and remark 4.6.2 in details.

Excercise 133 Prove Lemma 1.3 in [8] and deduce that for a group G and normal subgroup N the groups G and G/N are isoclinic if and only if $N \cap G' = 1$ is true. How can we use this theorem for the structure of the center of the radical? Compare the result to corollary 4.6.4. Which subgroups of G are isoclinic to G?

Excercise 134 Let G be a finite group. Prove a theorem of Hall that the breadth (which is the maximal length of all conjugacy classes of G) $b(G)$ and the degree of commutativity $\frac{c(G)}{|G|}$ are invariant under isoclinism. Why is the latter the probability that two randomly chosen elements within G are commutating? Compute both invariants for an extra-special group and evaluate the mentioned probability for $p \to \infty$ and for $|G| \to \infty$.

Excercise 135 Let G be an Abelian p-group of exponent p^e, for all $i \in \underline{e}$ let n_i be the number of cyclic subgroups of order p^{n_i} in a direct composition of G into cyclic groups and K a field of characteristic p. Based on theorem 4.2.1.4 we can calculate the invariants of $rad(KG)^\star$ by $s_1 Z_p \times \cdots \times s_e Z_{p^e}$ where for all $i \in \underline{e}$ let $s_i := k(|G^{p^{i-1}}| - 2 \cdot |G^{p^i}| + |G^{p^{i+1}}|)$. Prove the following facts:

(i) For all $i \in \underline{e}$ the order of G^{p^i} is exactly $p^{n_{i+1}} \cdots (p^{e-i})^{n_e}$.

(ii) Deduce that the invariants $s_1, \ldots, s_e$ can be determined by the invariants $n_1, \ldots, n_e, k$ and p.

(iii) By using $s_e = k(|G^{p^{e-1}}|-1)$ deduce $n_e = log_p(\frac{s_e}{k}+1)$.

(iv) By using $s_e + s_{e-1} = k(|G^{p^{e-2}}|-|G^{p^{e-1}}|) = k(p^{n_{e-1}} \cdot p^{2n_e} - p^{n_e})$ deduce $n_{e-1} = log_p(\frac{s_e+s_{e-1}}{kp^{n_e}}+1) - n_e$.

(v) By using $s_i + \cdots + s_e = k(|G^{p^{i-1}}|-|G^{p^i}|)$ deduce $n_i = log_p(\frac{s_e+s_{e-1}+\ldots s_i}{k|G^{p^i}|}+1) - n_e - \cdots - n_{i+1}$ for all $i \in \underline{e}$.

(vi) Deduce that the invariants of G can be determined recursively by the invariants of $rad(KG)^\star$. Reprove a theorem of Deskins stated in exercise 137.

(vii) Do a research in the literature (see [73]) for the following theorem: If the Abelian group G is a direct composition of cyclic groups $Z_1, \ldots, Z_n$ of order p^{r_i}, then the nilpotency class of $rad(KG)$ is exactly $1+\sum_{i=1}^{n}(p^{n_i}-1)$.

(viii) Let $G_2 = 1 + rad(KG)$. Prove that $cl(rad(KG)) = cl(rad(KG_2))$ if and only if $G = 1 + rad(KG)$ is valid. This is only the case for $|G|=|K|=2$.

Excercise 136 Let K be a field and G, H finite groups. True or false: $(KG)^\circ$ and $(KH)^\circ$ are isomorphic as Lie algebras if and only if G, H are of the same order.

Excercise 137 Prove the following theorem of Deskins (see [22]): Let K be a finite field of characteristic p and G, H finite Abelian p-groups. The group algebras KG and KH are isomorphic as associative algebras if and only if G, H are isomorphic as groups.

Excercise 138 Within exercise 137 analyze whether the mentioned statements are equivalent to:

(i) $E(KG)$ and $E(KH)$ are isomorphic as groups.

(ii) $1 + rad(KG)$ and $1 + rad(KH)$ are isomorphic as groups.

(iii) $rad(KG)^\star$ and $rad(KH)^\star$ are isomorphic as groups.

Excercise 139 Let K be a finite field of characteristic 2. For

(a) D_{32}

The invariants of the center

(b) an extra special 2-group of order 32

(c) $(D_8)^5$

(d) $(C_2)^5$

analyze for each pair G, H of groups whether

  (i) G and H are isomorphic as groups,

  (ii) $Z(G)$ and $Z(H)$ are isomorphic as groups,

  (iii) $Z(1+rad(KG))$ and $Z(1+rad(KH))$ are isomorphic as groups and/or

  (iv) G, H are isoclinic.

Excercise 140 Let K be a finite field of characteristic $p \neq 2$ and $n \in \mathbb{N}$. For

(a) an extra-special group of order p^{2n+1}

(b) an n-fold direct product of a non-Abelian p-group of order p^3

(c) an n-fold direct product of an Abelian p-group of order p^3

(d) $(Z_p)^n$

analyze for each pair G, H of groups of the same order whether

  (i) G and H are isomorphic as groups,

  (ii) $Z(G)$ and $Z(H)$ are isomorphic as groups,

  (iii) $Z(1+rad(KG))$ and $Z(1+rad(KH))$ are isomorphic as groups and/or

  (iv) G, H are isoclinic.

Excercise 141 Let $n \in \mathbb{N}$, p be a prime number, G be a p-group and K a field of characteristic p. Prove or disprove:

  (i) $Z(1+rad(KG))^{p^n} \cap G = Z(G)^{p^n}$

  (ii) $Z(1+rad(KG))^{p^n} \cap G = G^{p^n}$.

Excercise 142 Let p be a prime number and G, H be p-groups. Analyze the following topics:

  (i) If G, H are Abelian, then determine on what terms $(KG)^\circ$ and $(KH)^\circ$ are isomorphic as Lie algebras.

  (ii) If G, H are non-Abelian, then determine on what terms $Z(KG)^\circ$ and $Z(KH)^\circ$ are isomorphic as Lie algebras.

Excercise 143 Let $n \in \mathbb{N}$, p be a prime number and G be a non-Abelian p-group of order p^n. Prove that the derived subgroup is of order p if the center of G is of order p^{n-2}. Is the opposite implication true, too?

Excercise 144 Let $n \in \mathbb{N}$, p be a prime number and G be a non-Abelian p-group of order p^n. True or false: The class number is no power of p.

Excercise 145 Prove remark 4.2.2.4 in details.

Excercise 146 Execute example 4.1.1 for Q_{16}.

Excercise 147 Execute example 4.1.1 for SD_{16}.

Excercise 148 Execute example 4.1.1 for a p-group of order p^3.

Excercise 149 Do a research in the literature for the classification of p-groups of order p^4. Execute example 4.1.1 for every isomorphism class.

Excercise 150 Use exercise 149. For every representative of an isomorphic class of a group of order p^4 determine the exponent and the invariants of $Z(rad(KP))^*$. For this, let K be a field of order p^k. What is the effect of the group and of the field on the exponent? Do other factors exist which influence the determination of this exponent? What are the answers for the invariants?

Excercise 151 Apply theorem 4.1.4 to every subgroup of Q_{16}.

Excercise 152 Apply theorem 4.1.4 to every subgroup of D_{16}.

Excercise 153 Apply theorem 4.1.4 to every subgroup of SD_{16}.

Excercise 154 Let p be a prime number, G a p-group and K a finite field of characteristic p. What is the importance of the structure of unit groups of commutative group algebras for the structure of $Z(rad(KG))^*$?

Excercise 155 Within definition 4.1.6 determine the mentioned K-spaces and their dimensions for a group of order 3^3 and the field $GF(3^2)$.

Excercise 156 Let p be a prime number, G an Abelian p-group and K a finite field of characteristic p. Determine the invariants of $rad(KG)^*$ for the following cases:

(i) $p = 2, 3, 5$, $G = Z_p$, $K = GF(p)$

(ii) $p = 3$, $G = Z_3 \times Z_9 \times Z_{81}$, $K = GF(3^2)$

(iii) $p = 5$, $G = Z_{25} \times Z_{25}$

The invariants of the center

(iv) $p = 2$, $G = 7Z_2 \times 11Z_{2^4}$, $K = GF(4)$

(v) $p = 7$, $G = 7Z_7 \times 7Z_{7^2} \times 7Z_{7^3}$.

Excercise 157 Are the invariants for $rad(KG)^*$ within exercise 156 without gap, monotone or strict monotone? Determine the socle and the rank of $rad(KG)^*$.

Excercise 158 Within exercise 156 determine a complement of G in $rad(KG)^*$.

Excercise 159 Let p be a prime number, G, H Abelian p-groups and K a finite field of characteristic p. Is it possible to use the invariants of $rad(KG)^*$ and $rad(KH)^*$ to determine the invariants of $rad(K(G \times H))^*$? Is it possible to determine these invariants differently? What are the consequences for the case $H = G$?

Excercise 160 Use the results of exercise 159 to the groups within exercise 156 by forming direct products of two groups (which are not necessarily different).

Excercise 161 Let p be a prime number, $n, k, r \in \mathbb{N}$, G an Abelian p-group, K a finite field and $char(K) = p$. Determine the invariants of $rad(KG)^*$ in the following cases:

(i) $G = pZ_p \times p^2 Z_{p^2} \times \cdots \times p^n Z_{p^n}$, $K = GF(p^k)$

(ii) $G = pZ_p \times p^2 Z_{p^2} \times \cdots \times p^n Z_{p^n}$, $K = GF(p^n)$

(iii) $G = pZ_p \times p^2 Z_{p^2} \times \cdots \times p^n Z_{p^n}$, $K = GF(p)$

(iv) $G = pZ_p \times pZ_{p^2} \times \cdots \times pZ_{p^n}$, $K = GF(p^k)$

(v) $G = pZ_p \times pZ_{p^2} \times \cdots \times pZ_{p^n}$, $K = GF(p^n)$

(vi) $G = pZ_p \times pZ_{p^2} \times \cdots \times pZ_{p^n}$, $K = GF(p)$

(vii) $G = p^r Z_{p^n}$, $K = GF(p^k)$

(viii) $G = p^p Z_{p^p}$, $K = GF(p^p)$

(ix) $G = rZ_{p^n}$, $K = GF(p^k)$.

Excercise 162 Let p be a prime number, G an Abelian p-group and K a finite field of characteristic p. Are the invariants of $rad(KG)^*$ always without gap, monotone or strict monotone?

Excercise 163 For $G \in \{D_8, Q_8, SD_{16}\}$ and $K = GF(2)$ determine a complement of $Z(G)$ in $Z(1 + rad(KG))$.

Excercise 164 *Prove theorem 4.2.3.2 in details.*

Excercise 165 *Let p be a prime number, G an Abelian p-group and K a finite field of characteristic p. We call $rad(KG)^*$ homogenous if all invariants (different from zero) are identical. Determine the relevant groups G and fields K. True or false: If $rad(KG)^*$ is homogenous, then G is homogenous and vice versa. Define complete homogenous to be homogenous and without gap. Solve the exercise again for complete homogenous groups.*

Excercise 166 *Let p be a prime number, G an Abelian p-group, N a normal subgroup of G, U a subgroup of G and K a finite field of characteristic p. We call $rad(KG)^*$ homogenous if all invariants (different from zero) are identical. True or false: If $rad(KG)^*$ is homogenous, then $rad(KU)^*$, $rad(KN)^*$, $rad(K(G/N))^*$ and $rad(K(G \times G))^*$ are homogenous. Define complete homogenous to be homogenous and without gap. Solve the exercise again for complete homogenous groups.*

Excercise 167 *Prove proposition 4.3.2.3 in details.*

Excercise 168 *Let $k \in \mathbb{N}$ and K a finite field or order 2^k. For the following 2-groups G determine the invariants of $Z(rad(KG)^*)$ by using the chain of Frattini subgroups and draw the class-graph:*

(i) $G = D_{64}$

(ii) $G = Q_{32}$

(iii) $G = SD_{16}$

(iv) $G = Z_{128}$.

What is the influence of the field K? Are the invariants of the center without gap, monotone or strict monotone? Determine the rank and the socle of the center.

Excercise 169 *Let $k \in \mathbb{N}$ and K a finite field or order 2^k. For the following 2-groups G determine the invariants of $Z(rad(KG)^*)$ by using the chain of socle subgroups and draw the class-graph:*

(i) $G = D_{64}$

(ii) $G = Q_{32}$

(iii) $G = SD_{16}$

(iv) $G = Z_{128}$.

The invariants of the center   129

What is the influence of the field K? Are the invariants of the center without gap, monotone or strict monotone? Determine the rank and the socle of the center.

Excercise 170 Let p be a prime number, G an Abelian p-group and K a finite field of characteristic p. In what way can we use a graph to determine the structure of $rad(KG)^*$? (Tip: visualize the p-power structure) Analyze if two p-groups are isomorphic if and only if their assigned graph is isomorphic!

Excercise 171 Let K be a field of order 2, $G \in \{Q_{16}, D_{32}, SD_{64}\}$, U a subgroup and N a normal subgroup of G. Determine the order of the following elements in $Z(rad(KG)^*)$:

(i) $\overline{G}$

(ii) $\overline{U}$

(iii) $\overline{N}$

(iv) $\overline{c}$, c a conjugacy class of G

(v) $\overline{g^G}$, g an involution of G

(vi) $\overline{\bigcup_{c \in \mathcal{K}(G), |c|=2} c}$

(vii) $\overline{\bigcup_{c \in \mathcal{K}(G), |c|=4} c}$

(viii) $\overline{\bigcup_{c \in \mathcal{K}(G), |c|=8} c}$

(ix) $\overline{\bigcup_{c \in \mathcal{K}(G), |c|=16} c}$

(x) $\overline{\bigcup_{c \in \mathcal{K}(G), |c| \leq 2} c}$

(xi) $\overline{\bigcup_{c \in \mathcal{K}(G), |c| \leq 4} c}$

(xii) $\overline{\bigcup_{c \in \mathcal{K}(G), |c| \leq 8} c}$

(xiii) $\overline{\bigcup_{c \in \mathcal{K}(G), |c| \leq 16} c}$

(xiv) $\overline{\bigcup_{c \in \mathcal{K}(G), |c| \geq 2} c}$

(xv) $\overline{\bigcup_{c \in \mathcal{K}(G), |c| \geq 4} c}$

(xvi) $\overline{\bigcup_{c\in\mathcal{K}(G),|c|\geq 8} c}$

(xvii) $\overline{\bigcup_{c\in\mathcal{K}(G),|c|\geq 16} c}$

(xviii) $\overline{\bigcup_{c\in\mathcal{K}(G),|c|< 2} c}$

(xix) $\overline{\bigcup_{c\in\mathcal{K}(G),|c|< 4} c}$

(xx) $\overline{\bigcup_{c\in\mathcal{K}(G),|c|< 8} c}$

(xxi) $\overline{\bigcup_{c\in\mathcal{K}(G),|c|< 16} c}$

(xxii) $\overline{\bigcup_{c\in\mathcal{K}(G),|c|> 2} c}$

(xxiii) $\overline{\bigcup_{c\in\mathcal{K}(G),|c|> 4} c}$

(xxiv) $\overline{\bigcup_{c\in\mathcal{K}(G),|c|> 8} c}$

(xxv) $\overline{\bigcup_{c\in\mathcal{K}(G),|c|> 16} c}$.

In what way does the size of K influence these orders?

Excercise 172 Describe the class-graph for the generalized quaternion, the semi-dihedral and the dihedral 2-group of the same order for field of characteristic 2.

Excercise 173 Describe the class-graph for a field K of characteristic p and a p-group G such that the exponent of the center of G is $\frac{|G|}{p^2}$. Derive the exponent and the invariants of $\overline{\mathcal{K}(G)}^*$ from the class-graph.

Excercise 174 Let $K = GF(p)$. Describe the structure of the center of the radical of KG with respect to $*$ for the following p-groups:

(i) $G^3 \leq Z(G)$, $|G| = 3^9$

(ii) G is an extra-special p-group.

(iii) G is a special 5-group of order 5^{89}.

(iv) G is a minimal non-Abelian 7-group.

(v) The derived subgroup of G is of order 11 and the center of G is a cyclic subgroup of order 11^7.

(vi) The derived subgroup of G is exactly the center of G which is a cyclic group of order 17.

(vii) G is of order 17^3.

(viii) G is a regular 17-group of order 17^{35}.

(ix) $G \in \{D_{2^{11}}, D_{2^{12}}, SD_{2^{11}}, SD_{2^{12}}, Q_{2^{11}}, Q_{2^{12}}\}$

(x) $|G| = 5^{17}$ is valid, and the exponent of the center of G is $\frac{|G|}{5^2}$.

(xi) G is a 37-Sylow subgroup of $GL(3, 37)$

(xii) G is a 3-Sylow subgroup of $SL(37, 9)$

(xiii) G is a 5-Sylow subgroup of $PGL(3, 5)$

(xiv) G is a 17-Sylow subgroup of $PSL(8, 17)$

(xv) G is a non-Abelian 7-group of type 2 for which $r = 3, s = 5$ is valid

(xvi) G is a non-Abelian 2-group of type 3 for which $r = 3, s = 5$ is valid

(xvii) G is a non-Abelian 7-group of type 2 for which $r = 3, s = 5$ is valid

(xviii) G is an arbitrary p-Sylow subgroup of $S_3, A_3, S_4, A_4, S_5, A_5, S_6$

(xix) $G = Z_3 \wr Z_3$

(xx) $G = Z_2 \wr Z_2$

(xxi) $G = Z_p \wr Z_p$

(xxii) $G = Z_p \wr Z_p \wr \cdots \wr Z_p$

(xxiii) $G = Z_p \wr A$, A an Abelian p-group

(xxiv) $G = Z_p \wr H$, H a p-group

(xxv) $G = E \wr H$, E an elementary-Abelian and H a p-group.

In what way does the size of K influence this description? For each item draw the class-graph (if possible)!

Excercise 175 Let p be a prime number, G, H p-groups and K a finite field of characteristic p. Find an approach to determine the invariants of $Z(rad(K(G \times H)))^*$. In what way do the invariants of $Z(rad(KG))^*$ and $Z(rad(KH))^*$ influence this approach?

Excercise 176 *Apply exercise 175 to the following groups:*

(i) $G \times G$

(ii) G^n, $n \in \mathbb{N}$

(iii) $Q_8 \times D_{16}$

(iv) $D_{16} \times SD_{32}$

(v) $P \times P$, P is of order p^3

(vi) $D_{16} \times Z_{2^4}$

(vii) $G \times H$, G or H is Abelian

Excercise 177 *Let p be a prime number, G a p-group and K a finite field of characteristic p such that every Abelian subgroup is cyclic. Use chapter 3 (and there the representatives given for each isomorphism class of such groups) and determine the invariants of $Z(rad(KG))^*$. Draw the class-graph for each representative.*

Excercise 178 *Let p be a prime number, G a p-group and K a finite field of characteristic p such that every Abelian normal subgroup is generated by two elements. Use chapter 3 (and there the representatives given for each isomorphism class of such groups) and determine the invariants of $Z(rad(KG))^*$. Draw the class-graph for each representative.*

Chapter 5

Consequences for special types of unit groups

5.1 Unit groups with cyclic derived subgroup

For the determination of unit groups with a cyclic derived subgroup we need some preliminary results. The first one describes the nilradical of a modular group algebra for a cyclic p-group.

5.1.1 Proposition (nilradical for cyclic groups)

Let p be a prime number, K a field, $char(K) = p$ and G a p-group. If $z \in G$ exists such that $G = \langle z \rangle_{\mathcal{G}}$ is true, then $rad(KG) = (z - 1_G)KG$ is valid.

Proof. Based on theorem 1.1.21 we deduce $z - 1_G \in rad(KG)$, and thus $(z-1_G)KG$ is contained in $rad(KG)$. Let $n \in \mathbb{N}$ such that $o(z) = p^n$ is true. We use theorem 1.1.21 to obtain $rad(KG) = \langle \{z^i - 1_G \mid i \in \underline{n-1}\} \rangle_K$. Let $i \in \underline{n-1}$. We use the geometric series and conclude $z^i - 1_G = (z-1_G) \sum_{k=0}^{i-1} z^k$. Therefor, for every $i \in \underline{n-1}$ the element $z^i - 1_G$ is contained in $(z-1_G)KG$. ◊

Within exercise 179 a generalization of this proposition is presented.

5.1.2 Remark

Let A be an associative unitary K-algebra and $a \in A$ such that $Aa = aA$ is valid. By using an induction argument we deduce that for all $n \in \mathbb{N}$ the identity $(Aa)^n = Aa^n = a^nA = (aA)^n$ is true. ◊

We use proposition 5.1.1 for determining the exponent of the Frattini and the derived subgroup in a special case:

5.1.3 Lemma (special exponents)

Let p be a prime number, K a field, $char(K) = p$ and G a p-group. The following statements are valid:

(i) If G' is cyclic, then $exp(G') = exp((rad(KG)^*)')$ is valid.
In particular, $exp((rad(KG)^*)) \leq exp(G') \cdot exp(G/G')$ is true.

(ii) If $\Phi(G)$ is cyclic and K is finite, then $\Phi(G)$ and $\Phi(rad(KG)^*)$ are of the same exponent.
In particular, $exp((rad(KG)^*)) \leq exp(\Phi(G)) \cdot p$ is valid.

Proof. Let $N \in \{G', \Phi(G)\}$ and $z \in N$ such that $N = \langle z \rangle_{\mathcal{G}}$ is valid. By using proposition 1.1.14 and theorem 1.1.21 we deduce $ker\, p_N = KG\, rad(KN) = rad(KN)\, KG \subseteq rad(KG)$. Proposition 5.1.1 leads to $ker\, p_N = (z-1_G)KG = KG(z-1_G)$. Because of remark 5.1.2 we deduce that for all $x \in ker\, p_N$ the identity $x^{o(z)} = 0_{KG}$ is true. Therefor, the group $(ker\, p_N)^*$ is – based on corollary 2.4.4 – at most of exponent $o(z)$. Proposition 1.1.14 lets us deduce that the factor group $rad(KG)^*/(ker\, p_N)^*$ is $\mathcal{G}$-isomorphic to a subgroup of $rad(K(G/N))^*$. In the case $N = G'$ the group $rad(K(G/N))^*$ is an Abelian group, and thus $(rad(KG)^*)' \subseteq ker\, p_N$ is valid. In the case $N = \Phi(G)$ is the exponent of the p-group $rad(K(G/N))^*$ – based on theorem 4.2.1.4 – exactly p. Therefor, the group $\Phi(rad(KG)^*)$ is contained in $ker\, p_N$.

The add-on is valid by using (i) and (ii) and remark 4.2.1.3 about commutative unit groups.⋄

5.1.4 Theorem (cyclic derivation)

Let p be a prime number, K a field, $char(K) = p$ and G a p-group. $(rad(KG)^*)'$ is cyclic if and only if G is Abelian.

Proof. Let $(rad(KG)^*)'$ be cyclic. $G' - 1_G = ((G - 1_G)^*)'$ is a subgroup of $(rad(KG)^*)'$, and thus it is cyclic, too. Based on lemma 5.1.3 we deduce that $G' - 1_G$ and $(rad(KG)^*)'$ are possessing the same exponent. Both groups are cyclic, and therefor they are identical. $G - 1_G$ is containing $(rad(KG)^*)'$, and hence $G - 1_G$ is a normal subgroup of $rad(KG)^*$. We use corollary 1.2.4 to finish the proof.⋄

For uneven primes and small fields we can prove theorem 5.1.4 differently by using the following argument:

5.1.5 Remark

Let p be a prime number, K a field of order p, $char(K) = p$ and G a p-group. Based on [20] the group $Z_p \wr Z_p$ is involved in $E(KG)$ (A subgroup

Consequences for special types of unit groups   135

U of $E(KG)$ and a normal subgroup N of U exist such that $Z_p \wr Z_p$ is $\mathcal{G}$-isomorphic to U/N.). The group $Z_p \wr Z_p$ is not regular and possesses for $p \neq 2$ no cyclic derived subgroup.◇

5.1.6 Corollary (metacyclic unit groups)

Let p be a prime number, K a field, $char(K) = p$ and G a p-group. $rad(KG)^*$ is meta-cyclic if and only if $\mid K \mid = \mid G \mid = 2$ is true.

Proof. The corollary is a consequence of theorem 5.1.4 and corollary 4.2.2.7.◇

5.1.7 Corollary (cyclic Frattini subgroup)

Let p be a prime number, K a field, $char(K) = p$ and G a p-group. $\Phi(rad(KG)^*)$ is cyclic if and only if G is elementary-Abelian or $\mid K \mid = 2$ is valid and an element $n \in \mathbb{N}_0$ exists such that $G \cong_{\mathcal{G}} Z_4 \times \underbrace{Z_2 \times \cdots \times Z_2}_{n-times}$ is true.

Proof. Based on theorem 5.1.4 the group $\Phi(rad(KG)^*)$ is cyclic if and only if G is Abelian and $(rad(KG)^*)^p$ is cyclic. We use remark 4.2.1.3 to deduce the equivalence of this statement to the fact that G is Abelian and $rad(KG^p)^*$ is cyclic. Corollary 4.2.2.7 is used to obtain the equivalence to the statement that G is elementary-Abelian or $\mid K \mid = 2 = \mid G^2 \mid$ is true.◇

5.1.8 Remark (meta-Abelian unit groups)

Let p be a prime number, K a field, $char(K) = p$ and G a non-Abelian p-group.
In [60] it is proven by A. Shalev that $rad(KG)^*$ is not meta-Abelian for $p \geq 5$. Some years later, D.B. Coleman and D.S. Passman have proven in [21] that for $p = 3$ resp. for $p = 2$ the group $rad(KG)^*$ is meta-Abelian if and only if $\mid G' \mid = 3$ resp. G' is central in G and is $\mathcal{G}$-isomorphic to a subgroup of $Z_2 \times Z_2$.
F. Levin und G. Rosenberger obtain the same result in [44] for the topic on what terms the Lie algebra $rad(KG)^\circ$ is meta-Abelian.
B. Amberg and Y. Sysak are analyzing this phenomenon in their article [2] and they prove: For a radical ring R the group R^* is meta-Abelian if and only if the Lie ring R° is meta-Abelian.
We present an alternative prove that $rad(KG)^\circ$ – and thus also $rad(KG)^*$ – is not meta-Abelian for $p \geq 5$:
Let $a, b \in G$ such that $ab \neq ba$ is valid. Because of $(a \circ b) \circ (a^{-1} \circ b) = 0_{KG}$ we deduce

(1) $aba^{-1}b + baba^{-1} + a^{-1}b^2a - ab^2a^{-1} - a^{-1}bab - ba^{-1}ba = 0_{KG}$.

Because of $ab \neq ba$ and $p \neq 2$ we deduce $ba^{-1}ba \neq a^{-1}b^2a$ and $ba^{-1}ba \neq baba^{-1}$.

Case 1: $ba^{-1}ba = aba^{-1}b$
Based on (1) we deduce

(2) $baba^{-1} + a^{-1}b^2a - ab^2a^{-1} - a^{-1}bab = 0_{KG}$.

Because of $ab \neq ba$ and $p \neq 2$ we derive $a^{-1}b^2a \notin \{ab^2a^{-1}, a^{-1}bab\}$. We use (2) to obtain $p = 2$ which is a contradiction.

Case 2: $ba^{-1}ba \neq aba^{-1}b$
In this case we use (1) and $ba^{-1}ba \notin \{a^{-1}b^2a, baba^{-1}, aba^{-1}b\}$ to deduce $p = 3$ which is a contradiction.⋄

We summarize the results of this section:

5.1.9 Corollary (not-theorem)

Let p be a prime number, K a finite field, $char(K) = p$ and G a non-Abelian p-group. The following statements are valid:

(i) The derived subgroup of $rad(KG)^*$ is not cyclic.

(ii) $rad(KG)^*$ is not meta-cyclic.

(iii) $rad(KG)^*$ possesses no cyclic maximal subgroup.

(iv) $rad(KG)^*$ is no dihedral, semi-dihedral or quaternion group.

(v) For $p \neq 2$ the group $rad(KG)^*$ is not regular.

(vi) The Frattini subgroup of $rad(KG)^*$ is not cyclic.

(vii) $rad(KG)^*$ is no extra-special group.

(viii) $rad(KG)^*$ is not meta-Abelian for $p \geq 5$.⋄

5.2 Unit groups with cyclic p-power subgroup

The following proposition can be proven straightforward by using an induction argument:

5.2.1 Proposition

Let A be an associative K-algebra and $x, y \in A$. The following statements are valid:

(i) $\forall n \in \mathbb{N} : x \circ \underbrace{y \circ \cdots \circ y}_{n-times} = \sum_{k=0}^{n} \binom{n}{k}_K (-1_K)^k y^k x y^{n-k}$.

(ii) If p is a prime number and $char(K) = p$, then
$x \circ \underbrace{y \circ \cdots \circ y}_{p-times} = xy^p - y^p x$ is valid. ⋄

The following theorem summarizes some statements for the group of units in the non-Abelian case and smallest possible value of the derived subgroup of G. The first six statements are based on results of Shalev, Knoche and Du. In particular, the theorem of Du connects the determination of the class of nilpotency to the one of the associated Lie algebra. The last statement is added by the author and is based on proposition 5.2.1. Theorem 5.2.2 is important for this section (but also for some analysis made later on). Within the exercises some more results concerning the class of nilpotency are presented which may be studied by the reader by a research in the literature, too.

Let L be a Lie algebra and $n \in \mathbb{N}$. L is called n-Engel, if for all $x, y \in L$ the identity $x \cdot \underbrace{y \cdot \cdots \cdot y}_{n-times} = 0$ is valid. Similar n-Engel groups are defined by using commutators: a group G is called n-Engel if for all $x, y \in G$ the identity $[x, \underbrace{y, \ldots, y}_{n-times}] = 1$ is true. Within radical algebras R Amberg and Sysak have proven in [3] that R^* is n-Engel as group if and only if R° is m-Engel. The parameter m resp. n depends only on n resp. m.

5.2.2 Theorem (minimal class of Lie nilpotency)

Let p be a prime number, K a finite field, $char(K) = p$ and G a non-Abelian p-group. The following statements are equivalent:

(i) $cl(rad(KG)^*) = p$

(ii) $cl(rad(KG)^\circ) = p$

(iii) For all $x, y \in rad(KG)$ the identity $x \circ \underbrace{y \circ \cdots \circ y}_{p-times} = 0_{KG}$ is valid.

 (p-Engel property)

(iv) For all x in $rad(KG)$ the element x^p is central in $rad(KG)$.

(v) $|G'| = p$

(vi) Every non-central conjugacy class of G is of length p.

(vii) $exp(rad(KG)^*/Z(rad(KG)^*)) = p$

Proof. The equivalence of (i) and (ii) is a consequence of [24], the one of (v) and (vi) is based on [36], the one of (i) and (v) is deductable by [62], the one of (iv) and (vii) is true based on corollary 2.4.4 and the one of (iii) and (iv) is based on part (ii) of proposition 5.2.1. A statement within [61] is exactly the implication from (iii) to (v) and, finally, the implication from (ii) to (iii) is straightforward to be proven.◇

Let us remark that J. Kurdics has proven for $p > 2$ within [38] that part (v) of theorem 5.2.2 – which is $|G'| = p$ – is equivalent to the fact that the unit group of KG is p-Engel.

5.2.3 Corollary

Let p be a prime number, K a finite field, $char(K) = p$ and G a non-Abelian p-group. If one of the statements of theorem 5.2.2 is valid, then the following is true:

(i) $exp(Z(rad(KG)^*)) = exp(Z(G))$

(ii) $exp(G) \mid exp(rad(KG)^*) \mid p \cdot exp(Z(G))$

(iii) For $H := rad(KG)^*$ the identity $exp(Z(rad(KH)^*)) = exp(Z(rad(KG)^*))$ is valid.

Proof. ad(i): Part (iv) of theorem 5.2.2 lets us deduce that G^p is central in G. Thus, part (i) is a consequence of theorem 2.4.8 and proposition 2.4.7.

ad(ii): This part is deductable by part (i) and part (vii) of theorem 5.2.2.

ad(iii): $(rad(KG)^*)^p$ is – based on theorem 5.2.2 – central in $rad(KG)^*$. Thus, part (iii) is a consequence of theorem 2.4.8 and proposition 2.4.7.◇

5.2.4 Corollary

Let K be a field, $char(K) = 2$ and G a non-Abelian 2-group such that $|G'| = 2$ is valid and $Z(G)$ is elementary-Abelian. The identity $exp(rad(KG)^*) = 4$ is true.

Proof. The corollary is a direct consequence of corollary 5.2.3.◇

If G is a group resp. L a Lie algebra, then for all $n \in \mathbb{N}$ the set $Z_n(G)$

Consequences for special types of unit groups   139

resp. $Z_n(L)$ is called the n-th center of G resp. of L. The following theorem is proven in the case of a group algebra within the article [9] of A.A. Bovdi. Again, the theorem of Du is the main tool within its proof.

5.2.5 Theorem (elementary-Abelian factors alongside the upper central chain)

Let A be a K-radical algebra, p a prime number and $char(K) = p$. For all $n \in \mathbb{N}$ the statement $exp(Z_{n+1}(A^*)/Z_n(A^*)) = p$ is valid.

Proof. Based on [24] for all $n \in \mathbb{N}$ the sets $Z_n(A^*)$ and $Z_n(A^\circ)$ are identical. Based on corollary 2.4.4 for all $a \in A$ the statement $\underbrace{a * \cdots * a}_{p-times} = a^p$ is valid. The proof is finished by using proposition 5.2.1.◇

In our context of a modular group algebra all factor groups alongside the upper central chain – except the center itself – are elementary-Abelian. In their work [14] Bovdi and Milies extend this theorem to every normal subgroup of $1 + rad(KG)$. The size of these factor groups is not known to the author. A first insight into this problem is given within the dissertation of M. Theede in [81]. In the following diagram let $c := cl(rad(KG)^*)$.

- $Z_c(rad(KG)^*) = rad(KG)^*$

  { elementary-p-Abelian }

- $Z_{c-1}(rad(KG)^*)$

  $\cdots$ { stepwise elementary-p-Abelian }

- $Z_2(rad(KG)^*)$

  { elementary-p-Abelian }

- $Z(rad(KG)^*)$

  { structure see chapter 4 }

- $\{1\}$

The main insight for proving theorem 5.2.5 is that for a radical algebra A – based on the theorem of Du ([24]) – the upper central chains of A^* and A° are identical in every step. An analogue theorem is not valid for the descending central chains:

5.2.6 Remark

Let p be a prime number, K a finite field, $char(K) = p$ and G a non-Abelian p-group. $(rad(KG)^*)' \neq rad(KG) \circ rad(KG)$ is valid.

Proof. $rad(KG) \circ rad(KG) = \langle \{g \circ h \mid g, h \in G\} \rangle_K$ is valid. Thus, a subset B of $\{g \circ h \mid g, h \in G\}$ exists such that B is a K-basis of $rad(KG) \circ rad(KG)$. The derived subgroup of $rad(KG)^*$ contains the set $G' - 1_G$. G is non-Abelian, and thus an element $z \in G'$ exists such that $z \neq 1_G$ is true. If $rad(KG)' = rad(KG) \circ rad(KG)$ would be valid, then $z - 1_G$ would be contained in $rad(KG) \circ rad(KG)$. Thus, $z - 1_G \in \langle B \rangle_K$ would be true. For all $g, h \in G$ such that $gh = 1_G$ the statement $g \circ h = 0_{KG} \notin B$ is true which is a contradiction.⋄

After this excursus (which will be used later on) we return or focus to the main topic of this section: the determination of unit groups possessing a cyclic p-power subgroup.

5.2.7 Remark

Let p be a prime number, K a field and $char(K) = p$. For all $k, l \in K$ we use $char(K) = p$ to deduce that $k^p = l^p$ is valid if and only if k and l are identical. In particular, $k^p = 1$ is valid for all $k \in K \setminus \{0_K\}$ if and only if K is of order 2.⋄

5.2.8 Lemma

Let p be a prime number, K a field, $char(K) = p$ and G a p-group. If $(rad(KG)^*)^p$ is cyclic, then $\mid K \mid = 2$ or $exp(G) \leq p$ is valid.

Proof. Let $exp(G) \geq p^2$. An element $g \in G$ exists such that $o(g) = p^2$ is valid. Because of $p = o(g^p) = o(g^p - 1_G) = o((g - 1_G)^p)$ (see corollary 2.4.4) the subgroup $\langle g^p - 1_G \rangle_{\mathfrak{g}}$ is the unique subgroup of order p contained in $(rad(KG)^*)^p$. Let $k \in K \setminus \{0_K\}$. The identities $(k(g - 1_G))^p = k^p(g^p - 1_G) \neq 0_{KG}$ and $(k(g - 1_G))^{p^2} = 0_{KG}$ are valid, and thus we use corollary 2.4.4 to deduce $k^p(g^p - 1_G) \in \langle g^p - 1_G \rangle_{\mathfrak{g}}$. Hence, an element $i \in \underline{p-1}$ exists such that $k^p(g^p - 1_G) = g^{pi} - 1_G$ is true. Comparing coefficients leads to $k^p = 1_K$, and we finish the proof by using remark 5.2.7.⋄

5.2.9 Lemma

Let p be an uneven prime number, K a field, $char(K) = p$ and G a non-Abelian group of order p^3 and exponent p. Elements $x, y \in rad(KG)$ exists such that $dim_K \langle x^p, y^p \rangle_K = 2$ is valid.

Consequences for special types of unit groups 141

Proof. We execute the proof in several steps.
(1) Let us focus on the proof of theorem 14.10 on page 93 in [26]. Elements $a, b, c \in G$ exist such that $G = \langle a, b, c \rangle_\mathfrak{g}$, $b^a = bc$, $c^a = c$, $\langle b, c \rangle_\mathfrak{g} \cap \langle a \rangle_\mathfrak{g} = \langle c \rangle_\mathfrak{g} \cap \langle b \rangle_\mathfrak{g} = \{1_G\}$ and $\langle c \rangle_\mathfrak{g} = \Phi(G) = G' = Z(G)$ are valid. By using an induction argument we proof that for all $r, s \in \mathbb{N}$ the identity $b^r a^s = a^s b^r c^{sr}$ is valid. In addition, we remark that for every $g \in G$ exactly one triple $(i, j, k) \in \underline{p}^3$ exists such that $g = a^i b^j c^k$ is true.

(2) We define $x := a - b$ and $y = (a + b) - (1_G + c)$. Based on theorem 1.1.21 we deduce $x, y \in rad(KG)$, and by using $c \in Z(G)$ we obtain $x^p = (a - b)^p$ and $y^p = (a + b)^p - 2_K \cdot 1_G$.

(3) For all $g, h \in G$ and $n \in \mathbb{N}$ the identities

$$(g + h)^n = \sum_{k=0}^{n} \sum_{\substack{y_i \in \{g, h\} \\ |\{i | i \in \underline{n}, y_i = g\}| = k}} y_1 \ldots y_n \text{ and}$$

$$(g - h)^n = \sum_{k=0}^{n} \sum_{\substack{y_i \in \{g, h\} \\ |\{i | i \in \underline{n}, y_i = g\}| = k}} (-1_K)^{n-k} y_1 \ldots y_n \text{ are valid.}$$

(4) Because of (1) for every $k \in \underline{n}_0$ an element $z_k \in K\langle c \rangle_\mathfrak{g}$ exists such that

$$\sum_{\substack{y_i \in \{a, b\} \\ |\{i | i \in \underline{p}, y_i = a\}| = k}} y_1 \ldots y_p = a^k b^{p-k} z_k$$

is true. We use (3) and $exp(G) = p$ to deduce

$$(a - b)^p = \sum_{k=1}^{p-1} a^k b^{p-k} (-1_K)^{p-k} z_k \text{ and } (a + b)^p = \sum_{k=1}^{p-1} a^k b^{p-k} z_k.$$

(5) Now we prove $z_1 = \overline{\langle c \rangle_\mathfrak{g}}$ which leads to $z_1 \neq 0_{KG}$. If $i \in \underline{p-1}_0$, then we use (1) to deduce $b^i a b^{p-i-1} = a b^{p-1} c^i$, and the element z_1 is presentable as desired.

In addition, we prove z_2 is different from 0_{KG}. We have to analyse the sum

(5a) $$\sum_{\substack{y_i \in \{a, b\} \\ |\{i | i \in \underline{p}, y_i = a\}| = 2}} y_1 \ldots y_p = a^2 b^{p-2} z_2.$$

The sum contains $\frac{p(p-1)}{2}$ summands. We use (1) and derive $\sum_{i=0}^{p-2} a b^i a b^{p-2-i} = a^2 b^{p-2} \sum_{i=1}^{p-2} c^i$. The number of these summands is $p - 1$, and all of them are different. If $z_2 = 0_{KG}$ would be valid, then the sum in (5a) would possess – because of $char(K) = p$ – at least $p(p - 1)$ summands which is a contradiction.

(6) Now we prove the theorem for x and y from (2). Let us assume an element $l \in K$ exists such that $lx^p = y^p$ would be valid. Based on (4) we would derive

(6a) $\sum_{k=1}^{p-1} a^k b^{p-k} z_k (l(-1_K)^{p-k} - 1) = 0_{KG}$.

Because of $p \neq 2$ an element $k \in \underline{2}$ would exists such that $(-1_K)^{p-k} - 1 \neq 0_K$ would be valid. We use (5) and would derive the existence of an element $k \in \underline{2}$ such that $a^k b^{p-k} z_k (l(-1_K)^{p-k} - 1) \neq 0_{KG}$ would be true. In particular, elements $r \in \underline{p}$ and $t_r \in K$ would exist such that $a^k b^{p-k} t_r c^r (l(-1_K)^{p-k} - 1) \neq 0_{KG}$ would be true. Using (1) the „a, b, c-representation" of the elements of G is unique. Thus, the element $a^k b^{p-k} t_r c^r (l(-1_K)^{p-k} - 1)$ is different from every summand of the sum of (6a). Hence, the sum cannot be 0_{KG}.◇

5.2.10 Theorem (cyclic p-power subgroup, I)

Let p be a prime number, K a field, $char(K) = p$ and G a non-Abelian p-group. $(rad(KG)^*)^p$ is not cyclic.

Proof. For every group A the identity $A' \leq A^2$ is valid. Based on theorem 5.1.4 we can assume $p \neq 2$. In this case we prove the statement by an induction argument based on the order of G.
Let H be a proper subgroup of G. If $(rad(KG)^*)^p$ is cyclic, then $(rad(KH)^*)^p$ is cyclic, too. Thus, we can use induction to deduce that every maximal subgroup of G would be cyclic. The structure of these groups is well-known (see e.g. [26], page 309, exercise 22). Three cases are arising:

<u>Case 1:</u> G is $\mathcal{G}$-isomorphic to Q_8 which is a contradiction to $p \neq 2$.

<u>Case 2:</u> Elements $g, h \in G$ and $a, b \in \mathbb{N}$ exist such that $o(g) = p^a$, $o(h) = p^b$ and $|G| = p^{a+b}$ are valid. Because of $p \neq 2$ and lemma 5.2.8 we deduce $exp(G) = p$, and thus G is of order $|G| = p^2$ and Abelian which is a contradiction.

<u>Case 3:</u> Elements $g, h \in G$ and $a, b \in \mathbb{N}$ exist such that $o(g) = p^a$, $o(h) = p^b$ and $|G| = p^{a+b+1}$ are valid. Lemma 5.2.8 leads to $exp(G) = p$ and $|G| = p^3$. The derived subgroup of G is of order p, and we use theorem 5.2.2 to deduce that $(rad(KG)^*)^p$ is central in $rad(KG)^*$. Because of $|G| = p^3$ we obtain by using corollary 2.5.3 that $Z(rad(KG)^*)$ if of exponent p. $(rad(KG)^*)^p$ is cyclic, and thus lemma 5.2.9 leads to $|(rad(KG)^*)^p| = p$. Let $x \in rad(KG)$ such that $x^p \neq 0_{KG}$ is true. If $k \in K$ such that $k \neq 0_K$, then $(kx)^p = k^p x^p \neq 0_{KG}$ is valid. If $k, l \in K$ and $k^p x^p = l^p x^p$ are true, then $k^p = l^p$ and – because of $char(K) = p - k = l$ is valid. Thus, $|K| = p$

Consequences for special types of unit groups 143

and $(rad(KG)^*)^p = Kx^p$ are true (see corollary 2.4.4). In particular, $(rad(KG)^*)^p$ is containing no two-dimensional K-subspace contradicting lemma 5.2.9.⋄

5.2.11 Corollary (cyclic p-power subgroup, II)

Let p be a prime number, K a field, $char(K) = p$ and G a p-group. The following statements are equivalent:

(i) $(rad(KG)^*)^p$ is cyclic.

(ii) G is elementary-Abelian or G is Abelian, $p = 2$ and $|G^2| = |K^2| = 2$ are true.

Proof. The corollary is a consequence of theorem 5.2.10, proposition 1.1.14 and corollary 4.2.2.7.⋄

5.2.12 Corollary

Let p be a prime number, K a field, $char(K) = p$ and G a non-Abelian p-group. $exp(rad(KG)^*) \geq p^2$ is valid.

Proof. The corollary is a consequence of theorem 5.2.10 (because otherwise $(rad(KG)^*)^p$ would be the trivial group which is cyclic).⋄

This corollary shows us that lemma 5.1.3 is not valid for the subgroup of p-powers: If G^p is cyclic and K is finite, then G^p and $(rad(KG)^*)^p$ are not of the same exponent in general. To deduce this, let G be a non-Abelian p-group of exponent p. In this case G^p is trivial but $(rad(KG)^*)^p$ not because the exponent of $rad(KG)^*$ is at least p^2.

5.2.13 Corollary

Let p be a prime number, K a field, $char(K) = p$ and G a p-group. $rad(KG)^*$ is of exponent p if and only if G is elementary-Abelian.

Proof. The corollary is a consequence of corollary 5.2.12 and proposition 1.1.14.⋄

5.2.14 Corollary

Let p be a prime number, G a non-Abelian p-group possessing an elementary-Abelian center and K a field of characteristic p. If one of the statements of theorem 5.2.2 is valid, then the following items are true:

(i) $exp(Z(rad(KG)^*)) = p$

(ii) $exp(rad(KG)^*) = p^2$

(iii) For $H := rad(KG)^*$ the identity $exp(Z(rad(KH)^*)) = p$ is valid.

Proof. The corollary is a consequence of the corollaries 5.2.12 and 5.2.3.◇

5.2.15 Corollary

If p is a prime number, G a non-Abelian extra-special p-group and K a field of characteristic p, then $exp(rad(KG)^*) = p^2 = exp(G)$ is valid.

Proof. The corollary is a direct consequence of corollary 5.2.14. It can also be deduced by using corollary 5.2.12 and lemma 5.1.3 avoiding the deeper insight of theorem 5.2.2.◇

5.3 Unit groups for extra-special 2-groups

5.3.1 Remark

Let p be a prime number and G a p-group possessing a derived subgroup of order p. For all $g \in G \setminus Z(G)$ the identity $g^G = gG'$ is valid.◇

5.3.2 Lemma

Let K be a field, $char(K) = 2$, G a 2-group and $z \in G \setminus \{1_G\}$ such that $G' = Z(G) = \{1_G, z\}$ is valid. The following statements are valid:

(i) $KG(z + 1_G)$ is a zero-algebra.

(ii) $rad(KG) \circ rad(KG) = \langle\{\overline{g^G} \mid g \in G \setminus Z(G)\}\rangle_K$

(iii) $Z(rad(KG)) = (z + 1_G)KG = KG(z + 1_G) = \ker p_{Z(G)}$

(iv) $Z(rad(KG))$ is the $\mathcal{A}_1$-derivation of KG.

Proof. ad(i): We calculate $(z + 1_G)^2 = z^2 + 1_G = 1_G + 1_G = 0_{KG}$.

ad(ii): Let $g \in G \setminus Z(G)$. An element $h \in G$ exists such that $g \neq g^h$ is valid. Remark 5.3.1 leads to $g^G = \{g, g^h\}$. Because of $(h^{-1}g + 1_G) \circ (h + 1_G) = (h^{-1}g) \circ h = g^h + g$ and theorem 1.1.21 we deduce $\overline{g^G} \in rad(KG) \circ rad(KG)$. We use theorem 1.1.21 to obtain $rad(KG) \circ rad(KG) = \langle g \circ h \mid g, h \in G\rangle_K$. Let $g, h \in G$ and $gh \neq hg$. Based on $g^{-1}h^{-1}(g \circ h) = [g, h] + 1_G = z + 1_G$ we deduce $g \circ h = hgz + hg$. If hg is central in G, then $g \circ h = 0_{KG}$ is valid. In the other case we use remark 5.3.1 to deduce the identity $g \circ h = \overline{(hg)^G}$. Hence, $g \circ h \in \langle\{\overline{g^G} \mid g \in G \setminus Z(G)\}\rangle_K$ is true.

ad(iii),(iv): Propositions 5.1.1 and 1.1.14 lead to

Consequences for special types of unit groups   145

$\ker p_{Z(G)} = KG(KZ(G)(z+1_G)) = KG(z+1_G) = (z+1_G)KG$. Because of proposition 1.3.11 the identity $Z(rad(KG)) = Aug(KZ(G)) \oplus_K \langle\{\overline{g^G} \mid g \in G\setminus Z(G)\}\rangle_K$ is valid. $Aug(K(Z(G)) = K(z+1_G)$ is contained in $KG(z+1_G)$. Let $g \in G\setminus Z(G)$. We use remark 5.3.1 to obtain $\overline{g^G} = g + gz = g(z+1_G) \in KG(z+1_G)$. Thus, $Z(rad(KG)) \subseteq KG(z+1_G)$ is true. Let $g \in G$. We calculate $g(1_G + z) = g + gz$. If g is not central in G, then remark 5.3.1 leads to the identity $g + gz = \overline{g^G}$. In the other case we use $z^2 = 1_G$ to obtain the statement $g + gz = 1_G + z$. Therefor, $KG(1_G + z) \subseteq Z(rad(KG))$ is true, and thus we have proven $Z(rad(KG)) = (z+1_G)KG = KG(z+1_G) = \ker p_{Z(G)}$. In particular, $Z(rad(KG))$ is an ideal of KG. Because of (ii) we only need to deduce that $1_G + z$ is contained in the smallest ideal of KG containing $rad(KG) \circ rad(KG)$. If $g, h \in G$ such that $gh \neq hg$ is valid, then we use the identity $g^{-1}h^{-1}(g \circ h) = [g, h] + 1_G = z + 1_G$ to finish the proof.⋄

5.3.3 Theorem

Let K be a perfect field, $char(K) = 2$ and G an extra-special 2-group. The following statements are valid:

(i) $Z(rad(KG)^*) = (rad(KG)^*)^2$.

(ii) $Z(rad(KG)^*)$ is elementary-Abelian.

Proof. ad(i): Based on theorem 5.2.2 we deduce $(rad(KG)^*)^2 \subseteq Z(rad(KG)^*)$. Let $z \in G$ such that $G' = G^2 = Z(G) = \langle z \rangle_{\mathfrak{g}}$ is valid. Proposition 1.3.11 leads to $Z(rad(KG)^*) = K(z+1_G) \oplus_K \langle\{\overline{g^G} \mid g \in G\setminus Z(G)\}\rangle_K$. Let $k \in K$. An element $l \in K$ exists such that $l^2 = k$ is true. If $g \in G$ and $g^2 = z$, then

(1) $(l(1_G + g))^2 = l^2(1_G + z) = k(1_G + z) \in (rad(KG)^*)^2$

is valid. Let $a \in G \setminus Z(G)$. An element $b \in G$ exists such that $a \neq a^b$ is true, and based on remark 5.3.1 we deduce $a^G = \{a, a^b\} = \{a, az\}$. We use
$(l(h^{-1}g + h))^2 = l^2(h^{-1}gh + g + (h^{-1}g)^2 + h^2) = k\overline{g^G} + k(h^2 + (h^{-1}g)^2)$
and (1) and part (i) of lemma 5.3.2 to conclude (i).

ad(ii): This statement is a consequence of part (i) of lemma 5.3.2.⋄

5.3.4 Proposition

Let A be an associative unitary K-algebra and $a, b \in Q(A)$. The identity $[a, b] = (1_A + a' * b')(a \circ b)$ is valid.

Proof. Based on part (iii) of remark 1.1.4 we deduce $\{1_A + a, 1_A + b\} \subseteq E(A)$. By using the same remark again and $[1_A + a, 1_A + b] = (1_A + a)^{-1}(1_A +$

$b)^{-1}((1_A+a) \circ (1_A+b)) + 1_A$ we conclude $[a,b] = (1_A+a')(1_A+b')(a \circ b)$. Thus, the proof is finished.◇

5.3.5 Corollary

Let K be a field, $char(K) = 2$ and G an extra-special 2-group.

(i) $\forall a, b \in rad(KG) : [a,b] = (1_G + a*b)(a \circ b)$

(ii) $\forall a \in rad(KG), h \in G : [a, 1_G + h] = (1_G + a)h(a \circ h)$

Proof. Let $a, b \in rad(KG)$. Based on corollary 5.2.14 we deduce $exp(rad(KG)^*) = 4$, and thus $a^4 = b^4 = 0_{KG}$ is true. We use part (ii) of remark 1.1.16 to obtain $a' = a + a^2 + a^3$ und $b' = b + b^2 + b^3$. Lemma 5.3.2 and theorem 5.2.2 lets us deduce that $a \circ b$, a^2 and b^2 are central in $rad(KG)$. Now we use part (i) of lemma 5.3.2 to conclude that for all $x \in \{a^2, b^2, a^3, b^3\}$ the identity $x(a \circ b) = 0_{KG}$ is valid. Based on proposition 5.3.4 we deduce part (i), and part (ii) is a direct consequence of part (i).◇

5.3.6 Examples

Let K be a field, $char(K) = 2$, $k, l \in K$, G an extra-special 2-group, $z \in G \setminus \{1_G\}$, $Z(G) = \langle z \rangle_g$ and $g, h \in G$.

(i) By using remark 5.3.1 and corollary 5.3.5 we deduce

$[k(1_G + g), l(1_G + h)]$
$= (kl + k^2l + kl^2 + k^2l^2)\overline{(gh)^G} + (kl^2 + k^2l^2)\overline{g^G} + (k^2l + k^2l^2)\overline{h^G} + (k^2l^2)(1_G + z)$.

(ii) From (i) we obtain for $l = 1_K$ the identity

$[k(1_G + g), 1_G + h] = (k + k^2)\overline{g^G} + k^2(1_G + z)$.

(iii) In (i) the special case $k = l$ leads to

$[k(1_G + g), k(1_G + h)] = (k^2 + k^4)\overline{(gh)^G} + (k^3 + k^4)(\overline{g^G} + \overline{h^G}) + k^4(1_G + z)$.

(iv) If we apply (i) for k and l twice, then we obtain

$[k(1_G + g), l(1_G + h)] + [l(1_G + g), k(1_G + h)] = (k^2l + kl^2)(\overline{g^G} + \overline{h^G})$.

(v) Based on (iv) we deduce for $k \neq 0_K$ the identity

$[k^{-1}(1_G + g), k^2(1_G + h)] + [k^2(1_G + g), k^{-1}(1_G + h)] = (1_K + k^3)(\overline{g^G} + \overline{h^G})$.

(vi) By using remark 5.3.1 and corollary 5.3.5 we conclude

$[g + h, 1_G + h] = \overline{g^G} + \overline{(gh)^G} + 1_G + z.$

(vii) From (vi) we derive the identity

$[g + gh^{-1}, 1_G + gh^{-1}] = 1_G + z + \overline{g^G} + \overline{h^G}.$

5.3.7 Definition (a special subgroup)

For a group G and a field K we define

$U_{even} := \{x \mid \exists n \in \mathbb{N}_0, C_1, \ldots, C_{2n} \in \mathcal{K}(G) \setminus \{\{z\} \mid z \in Z(G)\}, x = \sum_{i=1}^{2n} \overline{C_i}\}.$

5.3.8 Theorem

Let K be a finite field, $char(K) = 2$ and G an extra-special 2-group. The following statements are valid:

(i) $(rad(KG)^*)' \leq Z(rad(KG)^*)$

(ii) $\forall k \in K, g, h \in G \setminus Z(G) : k(\overline{g^G} + \overline{h^G}) \in (rad(KG)^*)'$

(iii) The index of $(rad(KG)^*)'$ in $Z(rad(KG)^*)$ is at least $\frac{1}{2} \mid K \mid^2$.

(iv) If K is of order 2, then $(rad(KG)^*)' = Z(rad(KG)^*)$ or $(rad(KG)^*)' = \{0_{KG}, 1_G + z\} + U_{even}$ is valid.

Proof. ad(i): This statement is a direct consequence of theorem 5.2.2.

ad(ii): We remark that $(rad(KG)^*)'$ is – based on (i) and part (i) of lemma 5.3.2 – additive closed. Let $g, h \in G$.

Case 1: $gh \neq hg$
Let $k \in K$. We apply the identity in (ii) of example 5.3.6 to k^2 instead of k and gh instead of g and deduce

(1) $(k^2 + k^4)\overline{(gh)^G} + k^4(1_G + z) \in (rad(KG)^*)'.$

By addition of (1) to the equation in part (iii) of example 5.3.6 we obtain

(2) $(k^3 + k^4)(\overline{g^G} + \overline{h^G}) \in (rad(KG)^*)'.$

If we add the equations (v)-(vii) of example 5.3.6, then we deduce for $k \neq 0_K$ the statement

(3) $k^3(\overline{g^G} + \overline{h^G}) \in (rad(KG)^*)'$.

Parts (2) and (3) lead to $k^4(\overline{g^G} + \overline{h^G}) \in (rad(KG)^*)'$, and by using the perfectness of K we proof (ii).

<u>Case 2:</u> $gh = hg$
Let $k \in K$ and $x \in G \setminus (C_G(g) \cup C_G(h))$. We use case 1 to deduce that $k(\overline{x^G} + \overline{g^G})$ and $k(\overline{x^G} + \overline{h^G})$ are contained in $(rad(KG)^*)'$. The sum of these elements is $k(\overline{g^G} + \overline{h^G})$, and thus part (ii) is valid.

ad(iii),(iv): Both statements are a consequence of part (ii).◇

5.3.9 Proposition

Let p be a prime number, K a finite field, $char(K) = p$ and G a p-group such that $rad(KG)^*$ is a special p-group. G is an extra-special 2-group.

Proof. The class of nilpotency of $rad(KG)^*$ is exactly 2, and thus we use theorem 5.2.2 to deduce $p = 2$ and $|G'| = 2$. Because of $Z(rad(KG)^*) = (rad(KG)^*)^2 = (rad(KG)^*)'$ and the identities $G' = (1_G + rad(KG))' \cap G$ (see [19]) and $\Phi(G) = \Phi(1_G + rad(KG)) \cap G$ (see [49]) we conclude $Z(G) = G' = G^2$.◇

5.3.10 Example (non-special unit group)

Let K be a field of order 2, $G = D_8$ or $G = Q_8$ and $z \in Z(G) \setminus \{1_G\}$. The following statements are valid:

(i) $(rad(KG)^*)' = \{0_{KG}, 1_G + z\} + U_{even}$
   In particular, $1_G + rad(KG)$ is no special 2-group.

(ii) The normal closure of $G + 1_G$ in $rad(KG)^*$ is exactly $(G + 1_G) * (rad(KG)^*)'$.

Proof. ad(i): Based on proposition 1.3.11 the set $(G+1_G)*Z(rad(KG))^*$ is a normal subgroup of index 2 in $rad(KG)^*$. This normal subgroup is – by using corollary 1.3.7 – the normalizer of $G + 1_G$ in $rad(KG)^*$. Hence, for all $x \in rad(KG)^* \setminus ((G + 1_G) * (Z(rad(KG)^*))$ the group $rad(KG)^*$ is 𝒢-generated by $G + 1_G$ and x. Now we use part (i) of theorem 5.3.8 to deduce that the central derived subgroup is 𝒢-generated by $[1_G + g, 1_G + h]$ and $[1_G + g, x]$ $(g, h \in G)$. We can apply example 5.3.6 and theorem 5.3.8, and thus we only need to prove that for all $g \in G$ the element $[1_G + g, x]$ is contained in $\{0_{KG}, 1_G+z\}+U_{even}$. $(rad(KG)^*)'$ is – based on theorem 5.3.8 – central. Therefor, we only need to prove the statement for a 𝒢-generating set of G.

Case 1: $G = Q_8 = \{1_G, i^2, i, j, k, i^{-1}, j^{-1}, k^{-1}\} = \langle i, j \rangle_{\underline{g}}$:
Based on corollary 5.3.5 we calculate $[i+j, 1_G + i] = \overline{j^G} + \overline{k^G} + 1_G + z$ and $[i+j, 1_G+i] = \overline{i^G} + \overline{k^G} + 1_G + z$. For $x = i+j$ we deduce $x \notin N_{rad(KG)^*}(G+1_G)$ and $[x, 1_G + g] \in \{0_{KG}, 1_G + z\} + U_{even}$ for all $g \in G$.

Case 2: $G = D_8$:
Let h, a in G such that $G = \langle h, a \rangle_{\underline{g}}$, $o(a) = 2$, $o(h) = 4$ and $h^a = h^3$ are valid. Based on corollary 5.3.5 we deduce the identities $[a + h, 1_G + a] = \overline{h^G} + \overline{(ha)^G} + 1_G + z$ and $[a+h, 1_G+h] = \overline{a^G} + \overline{(ha)^G} + 1_G + z$. We use $x = a+h$ and obtain $x \notin N_{rad(KG)^*}(G+1_G)$ and $[x, 1_G + g] \in \{0_{KG}, 1_G + z\} + U_{even}$ for all $g \in G$.

ad(ii): Let N be the normal subgroup of $rad(KG)^*$ containing $G + 1_G$ and $g, h \in G$. Based on part (vii) of examples 5.3.6 we deduce for $gh \neq hg$ the statement

(1) $[g + gh^{-1}, 1_G + gh^{-1}] = 1_G + z + \overline{g^G} + \overline{h^G}$.

If $gh = hg$, then an element $x \in G \setminus (C_G(g) \cup C_G(h))$ exists. We apply (1) to x and g resp. to x and h. Thus, elements $a, b \in G$ and $r, s \in rad(KG)$ exists such that $[r, 1_G + a] + [s, 1_G + b] = \overline{g^G} + \overline{h^G}$ are valid. In addition, we calculate $[1_G + g, 1_G + h] = 1 + z$. Based on part (i) of theorem 5.3.8 we deduce that N is containing the set $\{0_{KG}, 1_G + z\} + U_{even}$, and by using (i) we finish the proof.◊

The following Hasse diagram presents several results proven within this work applied to $J := rad(GF(2)Q_8)^*$:

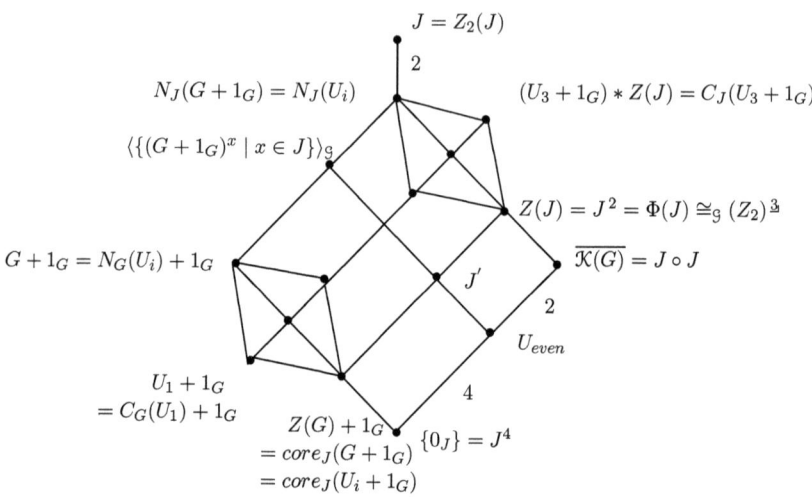

5.4 Open topics and exercises

Let n be a natural number, p be a prime number, G a finite p-group, K a field and $char(K) = p$.

Open-ended questions 5 *(i) Is the intersection of the p-powers of $1 + rad(KG)$ with G exactly G^p? Is the result true for all p^n-powers for all $n \in \mathbb{N}$?*

(ii) What is the size of the elementary-Abelian factor groups alongside the upper central chain of $rad(KG)^*$ (except the center itself)?

(iii) Are the factor groups alongside the descending central chain of $1 + rad(KG)$ elementary-Abelian? What is the structure of these Abelian groups?

(iv) Only for an extra-special 2-group the group $rad(KG)^*$ can be special. In the case $G = Q_8$ or $G = D_8$ and $K = GF(2)$ we have proven that $rad(KG)^*$ is not special. The author conjectures that $rad(KG)^*$ is never a special 2-group.

(v) What is the derived subgroup of $rad(KG)^*$? The case of a cyclic derivation of G might be a starting point of this analysis.

(vi) Is the normal closure of G in $1 + rad(KG)$ identical to $G \cdot (1 + rad(KG))'$? What is the normal closure of a subgroup or normal subgroup of G in $1 + rad(KG)$?

(vii) What is the chain of iterative commutator subgroups of $rad(KG)^*$ and the structure of the Abelian factor groups alongside this chain? On what term is a factor group elementary-Abelian? What is the length of this chain (which is the class of solvability)? What is the class of solvability for the associated Lie algebra $(KG)^\circ$ and for the associative algebra KG?

(viii) Determine those groups for which $(KG)^\circ$ resp. $E(KG)$ is n-Engel! Is the derived subgroup of G bounded in terms of n if $(KG)^\circ$ resp. $E(KG)$ is n-Engel?

(ix) Prove or disprove the equivalence for non-Abelian groups stated in exercise 179.

(x) Are the statements of theorem 5.2.2 equivalent for $p = 2$ to the fact that $1 + rad(KG)$ is 2-Engel? Under what conditions is $1 + rad(KG)$ exactly 2-Engel?

Excercise 179 *The aim of this exercise is to generalize the statement of proposition 5.1.1. The content of the proposition is: Let p be a prime number, K a field, $char(K) = p$ and G a p-group. If $z \in G$ exists such that $G = \langle z \rangle_g$ is true, then $rad(KG) = (z - 1_G)KG$ is valid. We describe some connections to this statements which are to be proven by the reader in more details as presented.*

In the case of a cyclic group the nilpotency class of $rad(KG)$ is identical to the nilpotency class of the element $z - 1$. This class is exactly the order of z ($(z-1)^n \neq 0$ for all $n \leq o(g) - 1$) which is the exponent of G and in this case the order of G. In addition, $rad(KG)$ is a principal ideal of KG. For every element $g \in G$ the class of $g - 1$ is exactly the order of g. For an Abelian group the exponent of G is exactly the maximum of all classes of elements of $rad(KG)$ (If G is Abelian and $a \in rad(KG)$, then $a^{exp(G)} = aug(a)^{exp(G)} \cdot 1_G$.) The following statements are of interest:

(i) G is cyclic.

(ii) $cl(rad(KG)) = |G|$

(iii) $cl(rad(KG)) = exp(G)$

(iv) $cl(rad(KG)) = max\{cl(a) \mid a \in rad(KG)\}$

(v) An element $g \in G$ exists such that $rad(KG) = (g-1)KG = KG(g-1)$.

(vi) An element $a \in rad(KG)$ exists such that $rad(KG) = KGa = aKG$.

(vii) Elements $a, b \in rad(KG)$ exist such that $rad(KG) = KGa = bKG$.

We want to prove the equivalence for Abelian p-groups at first. If G is cyclic, then (v) is valid based on proposition 5.1.1. The implications (v) $\longrightarrow$ (vi) $\longrightarrow$ (vii) are straightforward to be proven. If (vii) is valid, then $cl(rad(KG)) = cl(b)$ is true. Hence, statement (iv) is true. Based on our preliminary remarks $exp(G) = max\{cl(a) \mid a \in rad(KG)\}$ is true. Hence, (iii) and (iv) are equivalent for an Abelian group. Finally, the implications (iii) $\longrightarrow$ (ii) $\longrightarrow$ (i) are based on a result stated for Abelian p-groups in [73]: If the Abelian group G is a direct composition of cyclic groups $C_1, \ldots, C_n$ of order p^{r_i}, then the nilpotency class of $rad(KG)$ is exactly $1 + \sum_{i=1}^{n}(p^{n_i} - 1)$.

Now we turn to the non-Abelian case. Let $|G| = p^n$. Theorem 1 stated within [50] is exactly the equivalence of (i) and (ii): the nilpotency class is of maximal value $|G|$ if and only if G is cyclic. If G is cyclic, then we have already proven that $cl(rad(KG)) = exp(G)$ which is statement (iii). We want to prove the opposite implication in a way that for a non-Abelian p-group $cl(rad(KG)) > exp(G)$ is valid. We start withe case $p \neq 2$. Lets take an element $g \in G$ of maximal order $o(g) = p^e$. Shalev has proven within [64], proposition 3.2, that $cl(rad(KG)) \geq o(g) + (p-1)(n-e)$ is true. Because of $p \geq 3$ and $n > e$ this value is greater than p^e. The case $p = 2$ is handled by an induction argument. Theorem 2 stated within [50] lets us deduce that for a non-Abelian group G of order 8 the statement $exp(G) = 4 < cl(rad(KG)) < 8 = |G|$ is valid. We proceed by taking an element G of maximal order within G. If $\langle g \rangle_G$ is not a maximal subgroup of G, then we take a proper subgroup H of G such that $\langle g \rangle_G < H$ is true. Because of an induction argument we derive $cl(rad(KG)) \geq cl(rad(KH)) > exp(H) = exp(G)$. Thus, let $\langle g \rangle_G$ be a maximal (and therefore also normal) subgroup of index p. We can use again theorem 2 in [50] to prove that $cl(rad(KG)) > exp(G)$. This section shows us that also for a non-Abelian group the statements (i), (ii) and (iii) are equivalent.

If G is cyclic, then (v) is valid by using proposition 5.1.1. The implications $(v) \longrightarrow (vi) \longrightarrow (vii)$ are straightforward to be proven. The statements (vi) and (vii) are equivalent in a more general context (see Folgerung 11, page 69 in [76], a theorem of T. Nakayama). The implication from (vi) to (i) is stated in [35], page 299-300 (a theorem of Morita) in the context of a algebraic closed field of characteristic p. But a deeper insight in the proof lets us deduce that the cyclicity of G is be proven without using that the field is algebraic closed. Thus, (i), (v), (vi) and (vii) are equivalent.

By summarizing the results so far we haven proven also for a non-Abelian group that all statements except statement (iv) are equivalent. The implication from (iv) to (i) cannot be proven or disproven here and is left as an open topic in the previous section.

Excercise 180 Let $p \neq 2$. G is Abelian if and only if all elements of order p are central in G. For $p = 2$ the group G is Abelian if and only if all elements of order 2 and 4 are central in G. If needed, then do a research in the literature to prove this result. Are the mentioned statements equivalent to the fact that the mentioned elements form a subgroup or normal subgroup in G?

Excercise 181 On what terms is $rad(KG)^*$ cyclic?

Excercise 182 On what terms is $rad(KG)^*$ elementary-Abelian?

Excercise 183 On what terms is $Z(rad(KG)^*)$ elementary-Abelian?

Excercise 184 On what terms is $rad(KG)^*$ minimal non-Abelian?

Excercise 185 On what terms is $Z(rad(KG)^*)$ direct-indecomposable?

Excercise 186 *On what terms is $Z(rad(KG)^*)$ cyclic?*

Excercise 187 *On what terms is every Abelian normal subgroup of $rad(KG)^*$ cyclic? (Tip: consider the center)*

Excercise 188 *On what terms is every Abelian normal subgroup of $rad(KG)^*$ generated by two elements? (Tip: chapter 3, consider the center)*

Excercise 189 *Prove that the derived subgroup of G is exactly the intersection of G with the derived subgroup of $1 + rad(KG)$. Is the factor group of $1 + rad(KG)$ modulo its derived subgroup always elementary-Abelian?*

Excercise 190 *Determine a bound for the exponent of G by using the class of nilpotency of G and the exponent of the center of G. For this, use a result (proven within [26]) that the exponent of the center is maximal within the set of exponents of the factor groups alongside the ascending central chain of G. Apply this bound to $1 + rad(KG)$ and optimize this bound by using results proven within this work.*

Excercise 191 *Prove that the Frattini subgroup of G is exactly the intersection of G with the Frattini subgroup of $1 + rad(KG)$.*

Excercise 192 *Draw a Hasse diagram as done at the end of this chapter also for D_8, Q_{16}, D_{16}, SD_{16} and for $\mid G \mid = p^3$.*

Excercise 193 *Determine all consequences for $1 + rad(KG)$ which can be derived for groups with $\mid G \mid = p^4$ from the results within this work.*

Excercise 194 *Let G be cyclic. Determine $rad(KG)$ and $cl(rad(KG))$. Does an element x of $rad(KG)$ exist such that $cl(x) = cl(rad(KG))$ is true? Is the set of these elements a subspace or subgroup? Answer the questions in general and afterwards for Z_3, Z_9 and Z_{81} over $GF(3)$.*

Excercise 195 *Prove proposition 5.2.1 in details and apply it to Q_8, D_8 and $\mid G \mid = p^3$.*

Excercise 196 *Let $G := D_{16}$ and $K := GF(2)$. True or false:*

(i) The derivation of $rad(KG)^*$ is cyclic.

(ii) $rad(KG)^*$ ist meta-cyclic.

(iii) $rad(KG)^*$ possess a cyclic maximal subgroup.

(iv) $rad(KG)^*$ is a dihedral, semi-dihedral or quaternion group.

(v) $rad(KG)^*$ is regular.

Consequences for special types of unit groups   155

(vi) The Frattini subgroup of $rad(KG)^*$ is cyclic.

(vii) $rad(KG)^*$ is extra-special.

(viii) $rad(KG)^*$ is meta-Abelian.

(ix) $Z(rad(KG)^*)$ is Abelian.

(x) $Z(rad(KG)^*)$ is elementary-Abelian.

(xi) $Z(rad(KG)^*)$ is cyclic.

(xii) $Z(rad(KG)^*)$ is direct-decomposable.

(xiii) $(rad(KG)^*)^2$ is Abelian.

(xiv) $(rad(KG)^*)^2$ is cyclic.

(xv) $(rad(KG)^*)^2$ is elementary-Abelian.

(xvi) $Z(rad(KG)^*)$ possess the exponent 4.

(xvii) The invariants of $Z(rad(KG)^*)$ can be determined purely within G and K. If the answer is positive, then determine them.

(xviii) The normal closure of G in $1 + rad(KG)$ is $G \cdot (1 + rad(KG))'$.

Excercise 197 *Determine the nilradical of KG and its class of nilpotency for $G = Z_2 \times Z_2$ and $G = Z_4 \times Z_2$.*

Excercise 198 *Verify whether the following groups met the assumptions of theorem 5.2.2:*

(i) Q_8

(ii) D_8

(iii) $|G| = p^3$

(iv) $|G| = p^4$

(v) D_{16}

(vi) Q_{32}

(vii) If the assumptions are met, then they are met by every subgroup of G.

(viii) If the assumptions are met, then they are met by every normal subgroup of G.

(ix) If the assumptions are met, then they are met by every factor group of G.

(x) *If the assumptions are met, then they are met by $G \times G$.*

If the assumptions are not met, then analyze under what conditions they are met by the mentioned groups. In addition, verify the conclusions of the corollaries 5.2.3 and 5.2.4 if the assumptions are met.

Excercise 199 *On what terms is $Z_p \wr Z_p$ regular? On what terms does $Z_p \wr Z_p$ possess a cyclic derived subgroup?*

Excercise 200 *Determine the exponent of the derived subgroup of $rad(KG)^*$ for G' being of order 17.*

Excercise 201 *Determine the exponent of the derived subgroup of $rad(KG)^*$ if $G' = Z_{17^3}$ is valid.*

Excercise 202 *Determine the exponent of the Frattini subgroup of $rad(KG)^*$ if $\Phi(G)$ is of order 17.*

Excercise 203 *Determine the exponent of the Frattini subgroup of $rad(KG)^*$ if $\Phi(G) = Z_{17^3}$ is valid.*

Excercise 204 *On what terms is $rad(GF(2)Q_8)^*$ special?*

Excercise 205 *On what terms is $rad(GF(3)G)^*$ special or extra-special?*

Excercise 206 *On what terms is the exponent of $rad(KG)^*$ exactly 17?*

Excercise 207 *Prove remark 5.3.1 in details!*

Excercise 208 *Let N be a normal subgroup of G. Prove that $(1+(KG)Aug(KN)) \cap G = N$ is valid. Is this statement true for an arbitrary group G and field K?*

Excercise 209 *Let $p \neq 2$, $n \in \mathbb{N}$, $n \geq 4$, G a non-Abelian group of order p^n and of exponent p. Prove that elements $x, y \in rad(KG)$ exist such that $dim_K \langle x^p, y^p \rangle_K = 2$ is true.*

Excercise 210 *The aim of this exercise is to use the article [63] of Aner Shalev and the cited literature within it to prove and study the following statements for the class of nilpotency of $1 + rad(KG)$:*

(i) *Based on the theorem of Du the statement $cl(rad(KG)^*) = cl(rad(KG)^\circ)$ is true. Thus, the determination of the class of nilpotency of $E(KG)$ is transferred to the determination of the class of nilpotency of the associated Lie algebra $(KG)^\circ$.*

Consequences for special types of unit groups   157

(ii) $cl(E(KG)) = p$ is true if and only if the derived subgroup of G is cyclic (see theorem 5.2.2).

(iii) For $p \geq 5$ the statement $cl((KG)^\circ) = 1 + (p-1) \sum_{m \geq 1} m d_{(m+1)}$ is true in which $d_{(m+1)}$ is the index of the so-called dimension subgroups. This result leads to a systematic process of determining the class of nilpotency of $E(KG)$. (Jennings Lie theory of $(KG)^\circ$)

(iv) For $p \geq 5$ the statement $cl((KG)^\circ) \equiv 1 \mod p - 1$ is true. (consequence of Jennings Lie theory of $(KG)^\circ$)

(v) Let $n \in \mathbb{N}$, $|G'| = p^n$ and $p \geq 5$. $cl(E(KG)) \leq p^n$ is valid.

(vi) Let $n \in \mathbb{N}$, $|G'| = p^n$ and $p \geq 5$. $cl(E(KG)) = p^n$ is valid if and only if G' is cyclic.

(vii) Let $n \in \mathbb{N}$, $|G'| = p^n$ and $p \geq 5$. $cl(E(KG)) \geq (p-1)n + 1$ is true.

(viii) Let $n \in \mathbb{N}$, $|G'| = p^n$ and $p \geq 5$. $cl(E(KG)) = (p-1)n + 1$ is valid if and only if G' is elementary-Abelian and central.

(ix) Shalev describes those groups G for which the second and third lowest possible value of the class of nilpotency of $E(KG)$ is met.

Excercise 211 *The aim of this exercise is to study some aspects of the unitary subgroup. For this please use the article [16] of Bovdi and Rosa and the cited literature within it. The following topics are to be studied by the reader:*

(i) *What is the definition of the unitary subgroup?*

(ii) *What are the invariants of the unitary subgroup for an Abelian group G?*

(iii) *For $p \geq 3$ determine the order of the unitary subgroup.*

(iv) *On what terms is G a normal subgroup of the unitary subgroup?*

Chapter 6

A chain of p-groups

Within this chapter we focus on the chain of iterated p-groups defined by $G_0 := G$ and $G_{n+1} := 1 + rad(KG_n)$ for all $n \in \mathbb{N}$ over a finite field K of characteristic p and a non-Abelian p-group G. The previous chapters are linked to $G_1 = 1 + rad(KG)$. Now we want to study the behavior of this chain. Several parameters of this chain turn out to be increasing resp. unbounded (see proposition 6.1.2, e.g. the corresponding chain of derived subgroups, of breadth, of nilpotency classes, of strong derived length, of class numbers, of Baer-length). As a consequence the corresponding chain of degrees of commutativities converges against zero.

But the structure of the centers related to this chain can be described differently: the exponents are stable after the second step because the direct factor related to the class sums is elementary-p-Abelian. This result is generalized to arbitrary radical algebras (see theorem 6.2.11 and corollary 6.2.12). As a consequence we can prove that the chain of corresponding exponents and the chain of Engel-length of $(G_n)_{n \in \mathbb{N}_0}$ are unbounded (see theorem 6.3.3).

6.1 Basic properties

6.1.1 Definitions

Let p be a prime number, $G = G_0$ a non-Abelian finite p-group, K a finite field, $char(K) = p$ and $n \in \mathbb{N}$. We define $G_{n+1} := 1 + rad(KG_n)$ and call the chain $(G_n)_{n \in \mathbb{N}_0}$ the chain of iterated unit groups of modular group algebras.

Recall that the breadth of a p-group G is related to the maximum of all conjugacy class sizes $p^{b(G)}$. (For the definition and some results of the breadth see also [42] and [43].)

The derived length of a structure will be denoted by $st(\cdot)$. The structure can be a Lie algebra, a group or an associative algebra. The derived length is the length of the derived series related to the structure.

The degree of commutativity of a finite group is defined as the probability that two elements of the group commute. This value is discussed in details in [18]. One description of the degree of commutativity for a finite group G – denoted by $d(G)$ – is the value $\frac{c(G)}{|G|}$.

Finally, we introduce bounded Baer Lie algebras as done by Rips and Shalev within [54]. Let $n \in \mathbb{N}$. A Lie algebra L is called Baer bounded if all of its one-dimensional subalgebras are subnormal of bounded index resp. defect of subnormality. An equivalent definition is presented now: Given an element x of a Lie algebra L, define an n-stable series for $ad(x)$ as a decreasing sequence of Lie subalgebras $0 \subseteq L_n \subseteq L_{n-1} \subseteq \cdots \subseteq L_2 \subseteq L_1 = L$ satisfying:

(i) L_{i+1} is a Lie ideal of L_i for all $i \in \underline{n-1}$, and

(ii) $ad(x)$ acts trivially on L_i/L_{i+1} for all $i \in \underline{n-1}$.

We say that $ad(x)$ is n-Baer if $ad(x)$ has a n-stable series, and that L is Baer if every element of L is n-Baer for some natural number n. If this n is independent of x, we say that L is n-Baer. The minimal n is called the Baer-length of L. Obviously, n-Baer implies n-Engel. If L is n-Baer, we say that L is Baer-bounded. If L is nilpotent, then the Baer-length of L is bounded by the class of nilpotency of L.◊

6.1.2 Proposition (basic properties)

Let p be a prime number, G a non-Abelian finite p-group, K a finite field and $char(K) = p$. The chain $(G_n)_{n \in \mathbb{N}_0}$ possesses the following characteristics:

(i) The chain of p-groups $(G_n)_{n \in \mathbb{N}_0}$ is strict monotone increasing.

(ii) The chain of corresponding centers of $(G_n)_{n \in \mathbb{N}_0}$ is strict monotone increasing.

(iii) The chain of corresponding derived subgroups of $(G_n)_{n \in \mathbb{N}_0}$ is strict monotone increasing.

(iv) The chain of corresponding Frattini subgroups of $(G_n)_{n \in \mathbb{N}_0}$ is strict monotone increasing.

(v) The chain of corresponding breadth of $(G_n)_{n \in \mathbb{N}_0}$ is strict monotone increasing.

(vi) The chain of corresponding nilpotency classes of $(G_n)_{n \in \mathbb{N}_0}$ resp. of $((KG_n)^\circ)_{n \in \mathbb{N}}$ is strict monotone increasing for $n \geq 1$.

A chain of p-groups

(vii) The chain of corresponding associative nilpotency classes of $(rad(KG_n))_{n\in\mathbb{N}_0}$ is strict monotone increasing for $n \geq 1$.

(viii) The chain of corresponding p-power subgroups of $(G_n)_{n\in\mathbb{N}_0}$ is strict monotone increasing for $n \geq 2$.

(ix) The chain of corresponding associative or strong Lie derived lengths of $(KG_n)_{n\in\mathbb{N}_0}$ is unbounded.

(x) The chain of corresponding associative nilpotency classes of $(rad(K((G_n)')))_{n\in\mathbb{N}_0}$ is unbounded.

(xi) The chain of corresponding class numbers of $(G_n)_{n\in\mathbb{N}_0}$ is strict monotone increasing.

(xii) The chain of degrees of commutativity of $(G_n)_{n\in\mathbb{N}_0}$ converges against 0.

(xiii) The chain of Baer-length of $((KG_n)^\circ)_{n\in\mathbb{N}}$ is unbounded.

Proof. ad(i): Because of corollary 1.1.17 the chain $(G_n)_{n\in\mathbb{N}_0}$ is a chain of p-groups. This chain is strict monotone increasing because of remark 1.1.18.

ad(ii): This is a consequence of theorem 4.1.4.

ad(iii): By using (i) the chain is monotone increasing. Let us assume that an element $n \in \mathbb{N}$ exists such that $(G_{n-1})' = (G_n)'$ is true. In this case $(G_n)'$ is contained in G_{n-1}, and thus G_{n-1} is a normal subgroup of G_n. By using corollary 1.2.3 we derive that G_{n-1} is Abelian. Hence, G is Abelian which is a contradiction.

ad(iv): The argumentation is like presented in part (iii) because the Frattini subgroup contains the derived subgroup.

ad(v): Because of (i) the chain is monotone increasing. Let us assume that an element $n \in \mathbb{N}$ exists such that G_{n-1} and G_n have the same breadth. Let $g \in G_{n-1}$ such that the conjugacy class size of g is maximal in G_{n-1}. Then the classes of g in G_{n-1} and in G_n are identical because $g^{G_{n-1}} \subseteq g^{G_n}$ is true. The span of the class of g in G_{n-1} is normal in G_n. By using corollary 1.2.3 this span is central in G_{n-1}, and thus the breadth of G_{n-1} is zero which is a contradiction.

ad(vi): Because of (i) the chain is monotone increasing. We use a theorem of Bovdi and Kurdics in [15] that the nilpotency class of G and $1 + rad(KG)$ is identical only for $p = 2$ and G' is of order 2. Let us assume that an element $n \in \mathbb{N}$ exists such that $cl(G_n) = cl(G_{n+1})$ is true. We use the mentioned theorem to deduce that $p = 2$ and for all $i \in \underline{n}$ the identity $(G_i)' = (G_n)'$ would be valid. As done in part (iii) we obtain that G would be Abelian. The statement for the associated Lie algebra is derived by this result and the theorem of Du (see [24]) because the nilpotency classes of $1 + rad(KG_n)$ and $rad(KG_n)°$ are identical.

ad(vii): The associative nilpotency class is an upper bound for the Lie nilpotency class. Thus, statement (vii) is a direct consequence of part (vi).

ad(viii): Let us assume that an element $n \in \mathbb{N}_{\geq 2}$ exists such that G_{n-1} and G_n have the same p-power subgroup. Then, $(G_{n-1})^p = (G_n)^p$ is normal in G_n and contained in G_{n-1}. We use corollary 1.2.3 to deduce that $(G_{n-1})^p$ is central in G_n. Hence, $(G_n)^p$ is central in $1 + rad(KG_{n-1})$. We use theorem 5.2.2 to deduce that the derived subgroup of G_{n-1} is of order p and must coincide with the derived subgroup of G_{n-2} which is a contradiction to item (iii).

ad(ix) and (x): We use a theorem stated within [68] which bounds the associative derived length – which is also called strong Lie derived length – to the lower bound which is the upper integral part of the number

$$log_2(cl(rad(K(G_n)'))) + 1.$$

Thus, we have to prove that the associative nilpotency classes $cl(rad(K(G_n)'))$ tends to infinity. For this we use part (iii) and a bound presented in [34] on page 32: if the order of the derived group of a p-group G is p^n, then the associative nilpotence class of $rad(KG')$ is at least $n(p-1)$.

ad(xi): We use exercise 19 – which is based on the fixed point lemma – to deduce: if $(g_1)^G, \ldots, (g_c)^G$ are the conjugacy classes of G, then the conjugacy classes $(g_1)^{G_1}, \ldots, (g_c)^{G_1}$ are pairwise distinct. In particular, $c(G) \leq c(G_1)$ is valid. Let us assume that $c(G) = c(G_1)$ would be true. Then the classes of G_1 are exactly $(g_1)^{G_1}, \ldots, (g_c)^{G_1}$. The normal closure of G in G_1 is the smallest normal subgroup of G_1 containing G. The normal closure is contained in $(G_1)' \cdot G$. Therefor, the normal closure and $(G_1)' \cdot G$ would be identical to G_1. The intersection of $(G_1)'$ with G is – based on a theorem in [19] – exactly G'. Thus, the index of the derived subgroup of G and the one of G_1 would be the same. Let us focus on the linearization of the function $G \longrightarrow G/G'$, $g \mapsto gG'$. This function is an algebra epimorphism. The kernel is exactly $rad(KG') \cdot KG$. We conclude that the derived subgroup of $1 + rad(KG) = G_1$ is contained in $1 + rad(KG') \cdot KG$. We would derive that the order of G/G' is not smaller than $\mid K \mid^{\mid G/G' \mid -1}$. (For these facts see e.g. [35].) We prove that this is a contradiction. Let K be finite (otherwise we are done) and $k \in \mathbb{N}$ such that the order of K is p^k. Let $l \in \mathbb{N}$ such

A chain of p-groups

that the factor group G/G' is of order p^l. This factor group is not cyclic for a p-group (see e.g. [26]). We have to show that $p^{k(p^l-1)} > p^l$ is true. An easy induction argument leads to $2^n \geq n+2$ for all $n \in \mathbb{N}_{\geq 2}$. Thus, $p^{k(p^l-1)} \geq 2^{k(p^l-1)} > k(p^l-1) + 2 \geq p^l + 1 > p^l$.

ad(xii): We use two facts of the master thesis [18]: Proposition 3.1.1 lets us deduce that the corresponding sequence of degrees of commutativities is decreasing (because the degree of a subgroup is not smaller than the degree of the entire group). In addition, by proposition 2.1.10 the degree can be bounded by the upper bound $\frac{3}{2} \cdot \frac{1}{|g^G|}$ where g^G is of maximal size in G. In other words, $|g^G| = p^{b(G)}$. The corresponding breadths of the sequence are analyzed in part (v), and we know that the breadth-sequence is strict monotone increasing. Thus, we can bound our (monotone decreasing and convergent) sequence by another sequence which is convergent against zero.

ad(xiii): We use the quantitative version mentioned after theorems A' and B' within [54]: There is a function $h : \mathbb{N} \longrightarrow \mathbb{N}$ such that for all $n \in \mathbb{N}$ the following statement is valid: If $(KG)^\circ$ is n-Baer, then the order of the derived subgroup of G is bounded by $h(n)$. Thus, this item is now a consequence of item (iii).⋄

6.2 Bounded exponents of the centers

Now we focus our attention to the orders of the class sums with respect to the chain $(G_n)_{n \in \mathbb{N}_0}$.

6.2.1 Lemma

Let p be a prime number, G a non-Abelian finite p-group, K a finite field, $char(K) = p$ and $g \in G \setminus Z(G)$. The chain $(o(\overline{g^{G_n}}))_{n \in \mathbb{N}_0}$ of orders is monotone decreasing, bounded by p and convergent.

Proof. The proof is based on theorem 2.4.8. For the element g^G we know $o(\overline{g^G}) = p^{min\{n \in \mathbb{N} \mid C_G(g) < C_G(g^{p^n})\}}$. We use the same theorem to deduce for g^{G_1} that $o(\overline{g^{G_1}}) = p^{min\{n \in \mathbb{N} \mid C_{G_1}(g) < C_{G_1}(g^{p^n})\}}$ is valid. G is contained in G_1. Thus, if $C_{G_1}(g) = C_{G_1}(g^{p^n})$ is valid for an element $n \in \mathbb{N}$, then also – by intersecting with G – the identity $C_G(g) = C_G(g^{p^n})$ is true. Therefor, we have proven – based on theorem 2.4.8 – that $o(\overline{g^G}) \geq o(\overline{g^{G_1}})$ is valid, and the proof is finished.⋄

We want to determine the limit of this sequence and start with an example.

6.2.2 Example

We focus on $G := D_{16}$ and a finite field K of characteristic 2. Let $a, b \in G$ such that $o(b) = 2$, $o(a) = 8$, $a^b = a^{-1}$ and $G = \langle a, b \rangle_{\mathcal{G}}$ are valid. Based on example 3.1.2 we know $o(\overline{a^G}) = 4$. We use lemma 6.2.1 to deduce that $2 \leq o(\overline{a^{G_1}}) \leq 4$ is valid, and we prove that $\overline{a^{G_1}}$ is an involution. Based on theorem 2.4.8 we have to prove that $C_{G_1}(a) < C_{G_1}(a^2)$ is valid. Both centralizers can be described by using corollary 1.3.16: we have to calculate the orbits of G under $\langle a \rangle_{\mathcal{G}}$ resp. $\langle a^2 \rangle_{\mathcal{G}}$ by conjugation. The orbits of $\langle a \rangle_{\mathcal{G}}$ are $\{1\}$, $\{a\}$, $\{a^2\}$, $\{a^3\}$, $\{a^4\}$, $\{a^5\}$, $\{a^6\}$, $\{a^7\}$, $\{ab, a^3b, a^5b, a^7b\}$ and $\{b, a^2b, a^4b, a^6b\}$. Under $\langle a \rangle_{\mathcal{G}}$ the last two orbits are decomposing into four orbits of length 2: $\{ab, a^5b\}$, $\{a^3b, a^7b\}$, $\{a^2b, a^6b\}$ and $\{a^4b, b\}$. Thus, $C_{G_1}(a) < C_{G_1}(a^2)$ is valid and $\overline{a^{G_1}}$ is an involution.⋄

We extend the argumentation of this example and prove the following lemma (see [78], thanks to Arturo Magidin for his argument):

6.2.3 Lemma

Let p be a prime number, G a finite p-group and $g \in G$ such that $C_G(g) = C_G(g^p)$ is true. For all $x \in G \setminus C_G(g)$ the identity $x^{\langle g^p \rangle_{\mathcal{G}}} < x^{\langle g \rangle_{\mathcal{G}}}$ is valid.

Proof. (see [78]) Trivially, $x^{\langle g^p \rangle_{\mathcal{G}}} \subseteq x^{\langle g \rangle_{\mathcal{G}}}$. In order to get equality, we must have that $x^g \in x^{\langle g^p \rangle_{\mathcal{G}}}$, and hence that there exists k such that $g^{-1}xg = g^{-kp}xg^{kp}$. That would require $x = g^{-kp+1}xg^{kp-1}$, hence $x \in C_G(g^{kp-1}) = C_G(g) = C_G(g^p)$ (since $kp-1$ is relatively prime to p, and so g^{kp-1} generates the same cyclic subgroup as g). But that means that $x = x^g = x^{g^{kp}}$. In other words, if we have equality, then $x \in C_G(g)$. So if $x \notin C_G(g)$, then we cannot have equality of the two conjugacy orbits. In short, the equality will hold for all $x \in G \setminus C_G(g)$ if and only if that set is empty, if and only if $g \in Z(G)$.⋄

Alternatively, we can do a counting argument. Let $i \geq 0$ be the smallest nonnegative integer such that $g^{p^i} \in C_G(x)$. That is, $\langle g^{p^i} \rangle_{\mathcal{G}} = C_{\langle g \rangle_{\mathcal{G}}}(x)$. Then the number of distinct elements in the conjugacy orbit $x^{\langle g \rangle_{\mathcal{G}}}$ is p^i. If $i = 0$, then $x \in C_G(g)$, so we may assume that $i > 0$. But in that situation, $i - 1$ is the smallest nonnegative integer such that $(g^p)^{p^{i-1}} \in C_G(x)$, so that $\langle (g^p)^{p^{i-1}} \rangle_{\mathcal{G}} = C_{\langle g^p \rangle_{\mathcal{G}}}(x)$; hence the conjugacy orbit $x^{\langle g^p \rangle_{\mathcal{G}}}$ must have p^{i-1} elements. That is, in this situation, you cannot have equality. In particular, we also show that if $x \in G \setminus C_G(g)$, then $x^{\langle g \rangle_{\mathcal{G}}}$ has exactly p times as many elements as $x^{\langle g^p \rangle_{\mathcal{G}}}$ (see example 6.2.2).

Now we are ready to prove the following theorem:

A chain of p-groups 165

6.2.4 Theorem (decreasing property of G-classes)

Let p be a prime number, G a non-Abelian finite p-group, K a finite field, $char(K) = p$ and $g \in G \setminus Z(G)$. The following statements are valid:

(i) $o(\overline{g^{G_1}}) = p$

(ii) $\lim_{n \to \infty} (o(\overline{g^{G_n}}))_{n \in \mathbb{N}_0} = p$.

Proof. ad(i): If $C_G(g) < C_G(g^p)$ is true, then we use theorem 2.4.8 to deduce $o(\overline{g^G}) = p^{\min\{n \in \mathbb{N} \mid C_G(g) < C_G(g^{p^n})\}} = p$. By using lemma 6.2.1 we know that the sequence is decreasing and bounded by p. Thus, the sequence is constant with the value p. We turn to the case $C_G(g) = C_G(g^p)$. g is not central in G, and thus an element $x \in G \setminus C_G(g)$ exists. We use lemma 6.2.3 to deduce $x^{\langle g^p \rangle g} < x^{\langle g \rangle g}$. Therefor, the orbit of x under $\langle g \rangle_{\mathfrak{g}}$ decomposes under $\langle g^p \rangle_{\mathfrak{g}}$ into at least two orbits. By using corollary 1.3.16 we have proven that $C_{G_1}(g) < C_{G_1}(g^p)$ is valid. We finalize the proof by applying again theorem 2.4.8, but this time to $o(\overline{g^{G_1}}) = p^{\min\{n \in \mathbb{N} \mid C_{G_1}(g) < C_{G_1}(g^p)\}} = p$.

ad(ii): This part is a consequence of part (i) and lemma 6.2.1.⋄

The theorem shows us that certain elements of the center of $rad(KG)^*$ have a good behavior with respect to the chain $(G_n)_{n \in \mathbb{N}_0}$ of p-groups. It is very surprising that this is true for all non-central elements in $rad(KG)^*$ and can be proven within the context of finite radical algebras. We will present a brief introduction into the topic of radical algebras including the theorem of Du used earlier in this work (see e.g. theorem 5.2.2).

6.2.5 Definition and remark (radical algebras)

Let A be an associative K-algebra. By $rad(A)$ resp. $J(A)$ we symbolize the nilradical resp. the Jacobson radical of A. If A is right or left artian, then Gottfried Köthe has proven that both radicals coincide and are nilpotent. The associative nilpotency class is symbolized by $cl(A)$. A is called radical algebra, if $A = J(A)$ is valid. A is a nil algebra, if $A = rad(A)$ is true. $rad(A)$ and $J(A)$ are radical algebras.

The Lie algebra associated to A is symbolized by A° equipped with the multiplication $a \circ b := ab - ba$ for all $a, b \in A$. The upper central chain of A° is defined recursively by $Z_0(A^\circ) := \{0\}$ and $Z_n(A^\circ) := \{z \mid z \in A, \forall a \in A : z \circ a \in Z_{n-1}(A^\circ)\}$ for all $n \in \mathbb{N}$. A is Lie nilpotent, if a natural number $n \in \mathbb{N}$ exists such that $A = Z_n(A^\circ)$ is valid. The minimal n possessing this property is called the class of nilpotency of A° – symbolized by $cl(A^\circ)$. The lower central chain is defined recursively by $(A^\circ)^{(0)} := A$ and $(A^\circ)^{(n)} := (A^\circ)^{(n-1)} \circ A$ for all $n \in \mathbb{N}$. A is Lie nilpotent if and only if the lower central chains reaches the null space after finite many steps. The minimal number of these steps is again the class of nilpotency. Sufficient – and often used within applications – for the Lie nilpotency is the associative nilpotency, e.g. for $rad(A)$ if A is right artian. In [39] it is proven by Hartmut Laue that all members of the upper Lie central chain are associative subalgebras.

For all $a, b \in A$ we define (as original defined by Bartel Leendert van der Waerden) $a \star b := a + b + ab$, and we call $\star$ the circle or star composition on A. A is a monoid based on the composition $\star$, and 0 is its unit element. The group of units based on this monoid is called the star group or quasi regular group of A – symbolized by $Q(A)$. The elements of $Q(A)$ are called quasi regular, the inverse of $a \in Q(A)$ will be denoted by a^- or a', and the conjugated to a by b is symbolized by $a^{(b)} := b^- \star a \star b$. Every nilpotent element is quasi regular, and thus for every nil associative algebra A the identity $A = Q(A) = rad(A)$ is valid. If $Q(A) = A$ is true, then we use the symbol $A^\star$ for $Q(A)$. $rad(A)$ is a group based on $\star$. The Jacobson radical is a group based on $\star$, too. The upper central chain of $Q(A)$ is recursively defined by $Y_0(Q(A)) := \{0\}$ and $Y_n(Q(A)) := \{y \mid y \in Q(A), \forall a \in Q(A) : [y, a] \in Y_{n-1}(Q(A))\}$ for all $n \in \mathbb{N}$. The commutator $[y, a]$ is defined by $y^- \star y^{(a)}$ for all $y, a \in Q(A)$. $Q(A)$ is nilpotent, if the upper central chain of $Q(A)$ reaches $Q(A)$ in finite many steps. The minimal number of these steps is called the class of nilpotency of $Q(A)$ – symbolized by $cl(Q(A))$. The lower central chain of $Q(A)$ is defined recursively by $Q(A)^{(0)} = Q(A)$ and $Q(A)^{(n)} := [Q(A)^{(n-1)}, Q(A)]$ for all $n \in \mathbb{N}$. $Q(A)$ is nilpotent if and only if the lower central chain reaches the trivial subgroup after finite many steps. The minimal number of these steps is again the class of nilpotency. Within the literature $\gamma_k(G)$ is used for the k-th member of the lower central chain of a group G.

The group $Q(A)$ acts per conjugation on the additive group of A. For this operation another central chain can be defined recursively by $X_0(Q(A)) := \{0\}$ and $X_n(Q(A)) := \{x \mid x \in Q(A), \forall a \in Q(A) : x^{(a)} - x \in X_{n-1}(Q(A))\}$ for all $n \in \mathbb{N}$.

In addition, we define for a radical algebra A inductively $W_0(A) := \{0\}$ and $W_n(A) := \{w \mid w \in A, \forall a \in A : a^{(w)} - a \in W_{n-1}(A)\}$ for all $n \in \mathbb{N}$.⋄

Now we can present the results of Stephen Arthur Jennings, Hartmut Laue, Karsten Scholz and Xiankun Du. A first insight about a nilpotent connection between the group $A^\star$ the Lie algebra A° was proven by Stephen Arthur Jennings in 1955 (see [32]) for radical algebras:

6.2.6 Theorem (Jennings, 1955)

Let A be a radical algebra. $A^\star$ is nilpotent if and only if A° is nilpotent. ⋄

Stephen Arthur Jennings conjectured for radical algebras A that the nilpotency classes of $A^\star$ and A° coincide. In 1984 Hartmut Laue has proven some aspects of this conjecture in [40]:

6.2.7 Theorem (Laue, 1984)

Let A be an associative K-algebra. The following statements are valid:

(i) If $A = Q(A)$ and A is Lie-nilpotent, then $A^\star$ is nilpotent and $cl(A^\star) \leq cl(A^\circ)$ is valid.

(ii) Let A be nil, K a field and A° nilpotent. $Z_n(A^\circ) = Y_n(A^\star)$ is valid for all $n \in \mathbb{N}_0$.

(iii) If $Q(A) = A$ is valid, then $Z_2(A^\circ) = Y_2(A^\star)$ is true. ⋄

Hartmut Laue conjectured that for a radical algebra A the ascending central chains of $A^\star$ and A° are identical in every step. Xiankun Du has proven this conjecture in 1992 (see [24]). Thus, the conjecture of Stephen Arthur Jennings was proven, too.

6.2.8 Theorem (Du, 1992)

Let A be a radical algebra. For all $n \in \mathbb{N}_0$ the identity $Z_n(A^\circ) = Y_n(A^\star)$ is valid.

In particular, the conjecture of Jennings is true: $A^\star$ is nilpotent if and only if A° is nilpotent. If $A^\star$ or A° is nilpotent, then $cl(A^\circ) = cl(A^\star)$ is valid. ⋄

Further analysis performed by Karsten Scholz and Hartmut Laue (see [39] and [59]) yields to the following main theorem:

6.2.9 Main Theorem (Laue, Scholz, 1996)

Let A be a radical algebra. For all $n \in \mathbb{N}_0$ the identity

$$Z_n(A^\circ) = Y_n(A^\star) = X_n(A) = W_n(A)$$

is valid. In particular, every member of these four central chains are additive closed, associative subalgebras of A, Lie ideals of A° and normal subgroups of $A^\star$, and they are invariant under all automorphism of $A^\star$ and A° and under all $A^\star$-module automorphism of the additive group of A. ⋄

For proving the enhancement of theorem 6.2.4 we need the following elementary facts:

6.2.10 Proposition

Let K be field, V a K-space, $0 \neq \alpha$ an endomorphism of V, A an associative K-algebra and $a \in A$. The following statements are true:

(i) If α is nilpotent and $c := cl(\alpha)$, then $0 < ker(\alpha) < ker(\alpha^2) < \cdots < ker(\alpha^{c-1}) < ker(\alpha^c) = V$ is valid.

(ii) If $char(K) = p > 0$ is true, then $ad(a)^p = ad(a^p)$ is valid.

(iii) $ker(ad(a)) = C_A(a) = C_{A^\star}(a)$

Proof. ad(i): This is a basic fact for nilpotent endomorphism.

ad(ii): This statement is included in proposition 5.2.1

ad(iii): This is straightforward to verify.⋄

6.2.11 Theorem (conjugacy class sums in radical algebras)

Let K be a finite field of characteristic p and A a finite-dimensional K-radical algebra. For all $x \in A^\star \backslash Z(A^\star)$ the identity $o(\overline{x^{A^\star}}) = p$ is valid in $rad(KA^\star)^\star$. In particular, the direct factor $\overline{\mathcal{K}(A^\star)}$ of the center of $rad(KA^\star)^\star$ is elementary-p-Abelian.

Proof. K is finite and A is finite-dimensional. Thus, $A^\star$ is a p-group. P-groups are nilpotent, and thus A° is nilpotent based on theorem 6.2.6 of Jennings. Let $x \in A^\star \backslash Z(A^\star)$. Based on theorem 2.4.8 and corollary 2.4.4 we have to prove that $C_{A^\star}(x) < C_{A^\star}(x^p)$ is valid. (Let us remark that the p-th -power of x and the p-th $\star$-power of x are identical based on the mentioned corollary.) A° is nilpotent, and therefor $ad(x)$ is a nilpotent endomorphism of A°. Now we can use proposition 6.2.10 to finish the proof: $C_{A^\star}(x) = C_A(x) = ker(ad(x)) < ker(ad(x)^p) = ker(ad(x^p)) = C_A(x^p) = C_{A^\star}(x^p)$.⋄

This theorem – applied to finite radicals of associative algebras in modular characteristic – has influence on the chain of iterated unit groups of modular group algebras:

6.2.12 Corollary (chain of centers)

Let p be a prime number, G a non-Abelian finite p-group, K a finite field and $char(K) = p$. The chain $(G_n)_{n \in \mathbb{N}_0}$ possesses the following characteristics:

(i) The chain of exponents of $(ZG_n)_{n\in\mathbb{N}_{\geq 1}}$ is constant with the value $exp(Z(G_1)) = exp(Z(rad(KG)^\star))$.

(ii) Let $n \in \mathbb{N}_{\geq 2}$. The structure of center of G_n is the direct product of $1+rad(KZ(G_{n-1}))$ and the elementary-p-Abelian group $1+\mathcal{K}(G_{n-1})$. The latter one is of order $\mid K \mid^{c(G_{n-1})-\mid Z(G_{n-1})\mid}$.

(iii) Let $n \in \mathbb{N}_{\geq 2}$. The center of G_n is monotone decomposable and without gap.

(iv) Let $e \in \mathbb{N}$. If the chain of p^e-power subgroups of $(G_n)_{n\in\mathbb{N}_0}$ is bounded, then the chain of Engel-lengths of $((KG_n)^\circ)_{n\in\mathbb{N}_0}$ has this property, too.

(v) The chain of exponents of $(G_n)_{n\in\mathbb{N}_0}$ resp. of its central factor groups is bounded if and only if the chain of Engel-lengths of $((KG_n)^\circ)_{n\in\mathbb{N}_0}$ has this property, too.

(vi) The chain of Engel-lengths of $((KG_n)^\circ)_{n\in\mathbb{N}_0}$ is bounded if and only if the chain of Engel-lengths of $((rad(KG_n))^\star)_{n\in\mathbb{N}_0}$ possess this property.

Proof. ad(i)+(ii): These statements are a consequence of corollary 4.1.5 and theorem 6.2.11.

ad(iii): This statement is a consequence of (ii) and theorems 4.2.2.3 and 4.2.3.2.

ad(iv): Let $n \in \mathbb{N}$ such that $(G_n)^{p^e} = (G_{n+1})^{p^e}$ is valid. We use theorem 1.2.2 to deduce that $(G_{n+1})^{p^e}$ is central in G_n. Thus, for all $x, y \in rad(KG_n)$ the equation $x \circ y^{p^e} = 0$ is valid. We use proposition 6.2.10 to deduce (iv).

ad(v): If the chain of exponents is bounded, then elements $r, e \in \mathbb{N}$ exists such that the exponent for all groups G_x, $x \geq r$ is p^e. Let $x \geq r$ and $x, y \in G_x$. We deduce $y^{p^e} = 1$ and therefor $x \circ y^{p^e} = 0$. Now we apply proposition 6.2.10. For the opposite implication we use the argumentation before to deduce that the exponent of the factor groups modulo the center gets stable. The proof is finished by using part (i).

ad(vi): see the main theorem in [3]. ◊

Within the next remark we describe the structure of the centers related to the chain $(G_n)_{n\in\mathbb{N}_0}$.

6.2.13 Remark (invariants)

Let p be a prime number, G a non-Abelian finite p-group, K a finite field and $char(K) = p$. The chain of centers of $(G_n)_{n\in\mathbb{N}_0}$ is discussed within this remark.

The center of G_0 is the center of the group G.

The center of $G_1 = 1 + rad(KG)$ is discussed in details in chapter 3 and 4 of this work. It is the direct product of the Abelian group $1 + rad(KZ(G))$ and the cofactor $\overline{\mathcal{K}(G)}$.

In $Z(G_2)$ we need to calculate the Abelian factor $1 + rad(KZ(G_1))$ and the cofactor $1 + \overline{\mathcal{K}(G_1)}$. The second one is elementary-p-Abelian based on corollary 6.2.12, and for its calculation the number of conjugacy classes of G_1 is to be calculated which is not known to the author. For the first factor we need to calculate $1 + rad(K(A \times B))$ for two Abelian p-groups A, B.

In $Z(G_n)$ for $n \geq 3$ the situation is similar: we need to calculate the Abelian factor $1 + rad(KZ(G_{n-1}))$ and the cofactor $1 + \overline{\mathcal{K}(G_{n-1})}$. The second one is again elementary-p-Abelian, and for its calculation the number of conjugacy classes of G_{n-1} is to be calculated. For the first factor we need to calculate $1 + rad(K(A \times B))$ for two Abelian p-groups A, B such that B is elementary-p-Abelian.

We want to finalize this remark by demonstrating how to calculate $1 + rad(KZ(G_1))$ resp. $1 + rad(KZ(G_{n-1}))$. Let A, B two Abelian p-groups. For a start we assume that B is elementary-p-Abelian. For this we use theorem 4.2.1.4 to calculate the invariants. Let $\mid K \mid = p^k$. The invariants are $s_i := k(\mid (A \times B)^{p^{i-1}} \mid -2 \cdot \mid (A \times B)^{p^i} \mid + \mid (A \times B)^{p^{i+1}} \mid)$. Let $exp(A) = p^r$. B is elementary-p-Abelian, and thus the invariants of $1 + rad(KA)$ and $1 + rad(K(A \times B))$ are identical for $i = 2, \ldots, r$. The invariant s_1 is the invariant a_1 for $1 + rad(KA)$ plus $k(\mid A \mid (\mid B \mid -1))$.

Now let B be an arbitrary p-group. This case is also discussed within exercise 159 in chapter 4. We will give the reader an idea how to solve this exercise. We use the equation $\forall i \in \underline{e} : s_i + \cdots + s_e = k(\mid (A \times B)^{p^{i-1}} \mid - \mid (A \times B)^{p^i} \mid)$ of theorem 4.2.1.4. Let $exp(A) = p^a$ and $exp(B) = p^b$, and we assume w.l.o.g. that $a \geq b$ is valid. Let $a_1, \ldots, a_a$ resp. $b_1, \ldots, b_a$ the invariants of $1 + rad(KA)$ resp. $1 + rad(KB)$. If needed, then we set $b_{a+1} := \cdots := b_a := 0$. Let us focus on $s_a = k(\mid (A \times B)^{p^{a-1}} \mid -1) = k(\mid A^{p^{a-1}} \mid \cdot \mid B^{p^{a-1}} \mid -1)$. We can use the rule $ab - 1 = (a-1)b + (b-1)$ for any numbers. Thus, $s_a = a_a \cdot \mid B^{p^{a-1}} \mid + b_b$ is valid. We calculate one further step, and the rest an to be done explicitly as an exercise. The equation $s_a + s_{a-1} = k(\mid (A \times B)^{p^{a-2}} \mid - \mid (A \times B)^{p^{a-1}} \mid)$ is valid. In this case we use the rule $a_2 b_2 - a_1 b_1 = (a_2 - a_1) b_2 + a_1 (b_2 - b_1)$ for arbitrary numbers a_1, b_1, a_2, b_2. Thus, $s_a + s_{a-1} = a_{a-1} \mid B^{p^{a-2}} \mid + b_{a-1} \mid A^{p^{a-1}} \mid$ is valid. s_a was already determined, so we can deduce a rule for s_{a-1} and so on.$\diamond$

6.3 Unbounded exponents

Now we focus on the chain of exponents of $(G_n)_{n\in\mathbb{N}_0}$ and we want to prove that it is unbounded. The proof is done based on several lemmas. I want to say thank you to Mr. C. Baginski for the hint of Shalev's article [66].

6.3.1 Lemma

Let p be a prime number, G a non-Abelian finite p-group, K a finite field and $char(K) = p$. If the chain of exponents of $(G_n)_{n\in\mathbb{N}_0}$ is bounded, then the chain of derived length of $(G_n)_{n\in\mathbb{N}_0}$ is bounded.

Proof. We use the method of polynomial identities applied by Aner Shalev within example 6.1 in [66]. If $exp(1 + rad(KG)) = p^E$ is valid for a natural number E, then $x^{p^E} = 0$ is true for every element $x \in rad(KG)$. Therefore, the group algebra satisfy the polynomial identity

$$(x+y)^{p^E} = x^{p^E} + y^{p^E}$$

of degree p^E. By the theorem of Passman on polynomial identities in group rings of characteristic p (see [51] or [52]) the group G possesses a normal subgroup N such that

$$\mid G:N \mid \cdot \mid N' \mid$$

is bounded above by a fixed function on p^E. Consequently, if the sequence $(G_n)_{n\in\mathbb{N}_0}$ is satisfying

$$min\{\mid G_n : N \mid \cdot \mid N' \mid ; N \trianglelefteq G_n\} \to \infty \ as\ n \to \infty$$

, then

$$exp(1 + rad(KG_n)) \to \infty \ as\ n \to \infty.$$

Observe that $1 + rad(KG_n) = G_{n+1}$ is true for all $n \in \mathbb{N}$ per definition. So, if the exponents are bounded, then

$$(min\{\mid G_n : N \mid \cdot \mid N' \mid ; N \trianglelefteq G_n\})_{n\in\mathbb{N}}$$

must be bounded, too. We use this fact to deduce that the derived length are bounded, too.

Let the sequence

$$(min\{\mid G_n : N \mid \cdot \mid N' \mid ; N \trianglelefteq G_n\})_{n\in\mathbb{N}}$$

be bounded. First we note, that the sequence is increasing for increasing n. If N is a normal subgroup of G_{n+1}, then $N \cap G_n$ is normal in G_n. The index $\mid G_n : N \cap G_n \mid$ is identical to the index $\mid (G_n \cdot N) : N \mid$ which is not greater than $\mid G_{n+1} : N \mid$. It is identical, if and only if $G_{n+1} = G_n \cdot N$ is true. In addition, the order of $(N \cap G_n)'$ is not greater than the order of N'. It is identical if and only if $N' = (N \cap G_n)'$ is valid. Thus, the sequence of minima is increasing. If it is bounded, then it must be stabile. Therefore, an element $m \in \mathbb{N}$ exists such that

$$min\{\mid G_m : N \mid \cdot \mid N' \mid; N \trianglelefteq G_m\} = min\{\mid G_r : N \mid \cdot \mid N' \mid; N \trianglelefteq G_r\}$$

is valid for all $r \geq m$.

Let $r \geq m+1$. Take N normal in G_r such that $\mid G_r : N \mid \cdot \mid N' \mid$ is minimal. As before we focus on $N \cap G_m$. By the choice of N and stability of the minima we deduce

$$\mid G_m : N \cap G_m \mid \cdot \mid (N \cap G_m)' \mid = \mid G_r : N \mid \cdot \mid N' \mid.$$

We have already proven that under this conditions $G_r = G_m \cdot N$ and $N' = (N \cap G_m)' \subseteq G_m$ are valid. N' is characteristic in N which is normal in G_r. Therefore N' is normal in G_r contained in $G_m \subseteq G_{r-1}$. We use corollary 1.2.3 to deduce that N' is central in G_{r-1}. The center of G_{r-1} is contained in the center of G_r (see e.g. corollary 4.1.5). Thus, the normal subgroup N is of derived length at most 2. Now, $G_r = G_m \cdot N$ is of bounded derived length.⋄

The next lemma is identical to lemma 4.3 on page 93 in [54]:

6.3.2 Lemma

Let L be a Lie algebra and n, d be natural numbers. If L is n-Engel and of derived length d, then L is $(d \cdot n)$-Baer.⋄

Now we are ready to prove the theorem of unbounded exponents:

6.3.3 Theorem (unbounded exponents)

Let p be a prime number, G a non-Abelian finite p-group, K a finite field and $char(K) = p$. The chain of exponents of $(G_n)_{n \in \mathbb{N}_0}$ is unbounded.

Add-On: The chain of Engel-length of $((KG_n)^\circ)_{n \in \mathbb{N}_0}$, the chain of Engel-length of $(G_n)_{n \in \mathbb{N}_0}$, the exponents of the central factor groups of the chain $(G_n)_{n \in \mathbb{N}_0}$ and for all fixed $e \in \mathbb{N}$ the chain of p^e-power subgroups of $(G_n)_{n \in \mathbb{N}_0}$ are unbounded.

Proof. Let us assume that the chain of exponents of $(G_n)_{n \in \mathbb{N}_0}$ is bounded. We use part (v) of corollary 6.2.12 to deduce that the chain of Engel-length of the Lie algebras $((KG_n)^\circ)_{n \in \mathbb{N}_0}$ would be bounded, too.

In view of lemma 6.3.1 the chain of derived length of $(G_n)_{n \in \mathbb{N}_0}$ would be bounded, too. We use a theorem of Amberg and Sysak on radical algebras stated within [4]: If its adjoint group is of derived length n, then its Lie algebra is of derived length $m(n)$ for a fixed function m depending only on n. Therefore, the chain of derived length of the Lie algebras $((KG_n)^\circ)_{n \in \mathbb{N}_0}$ would be bounded, too.

Combining all, we would have proven that the derived length and the Engel-length of the Lie algebras $((KG_n)^\circ)_{n \in \mathbb{N}_0}$ would be bounded. In view of lemma 6.3.2 the Baer-length of the Lie algebras $((KG_n)^\circ)_{n \in \mathbb{N}_0}$ would be bounded. This is a contradiction to part (xiii) of proposition 6.1.2.

The add-ons are a direct consequence of corollary 6.2.12 and the theorem proven so far. ◇

6.3.4 Remark

(i) The chain of exponents might not be increasing strictly. A. Shalev has proven within [66] that the exponents of G and $1 + rad(KG)$ can be the same. In the article [11] A. Bovdi and P. Lakatos have proven similar results on the exponent of $1 + rad(KG)$ such that it can be very close to the one of G.

(ii) Let us focus again on the chain $(G_n)_{n \in \mathbb{N}_0}$ of p-groups. Let $n, r \in \mathbb{N}$. We take a group G_n and within G_n the r-th center $Z_r(G_n)$. We are able to bound the exponent of this center. In view of corollary 6.2.12 and theorem 5.2.2 the exponent of the r-th center is not greater than $exp(Z(G_1)) \cdot p^{r-1}$ which can be bounded by $exp(G)) \cdot p^{r-1}$. Therefore, the classes of nilpotency must not be bounded according to the chain. ◇

We summarize some of the properties of the chain $(G_n)_{n \in \mathbb{N}_0}$ of iterated p-groups. Let r be a fixed natural number:

structural value	property
$\mid G_n \mid$	$\to \infty$
$\mid (G_n)' \mid$	$\to \infty$
$\mid (G_n)^{p^r} \mid$	$\to \infty$
$\mid \Phi(G_n) \mid$	$\to \infty$
$b(G_n)$	$\to 0$
$cl(G_n)$	$\to \infty$
$cl((KG_n)^\circ)$	$\to \infty$
$cl(rad(KG_n))$	$\to \infty$
$c(G_n)$	$\to \infty$
G_n-Engel-length	$\to \infty$
$(KG_n)^\circ$-Engel-length	$\to \infty$
$(KG_n)^\circ$-Baer-length	$\to \infty$
$exp(G_n)$	$\to \infty$
$exp(G_n/Z(G_n))$	$\to \infty$
$exp(Z(G_n))$	$\to exp(Z(G_1)) \le exp(G)$
$exp(Z_r(G_n))$	is bounded by $exp(Z(G_1)) \cdot p^{r-1}$
$exp(Z_r(G_n))$	$\to$???

The following diagram visualize the behaviour of the center of the chain $(G_n)_{n\in\mathbb{N}_0}$ of iterated p-groups:

G_0
$C_{G_0}(g^{p^i})$
$C_{G_0}(g) = \ldots = C_{G_0}(g^{p^{a-1}})$
$Z(G_0)$

1_{G_0}

determines

$1 + rad(KZ(G_0))$

$\overline{K(G_0)}$
1_{G_0}

determined by conjugacy class sums and their power structure

$G_1 = 1 + rad(KG_0) = Z_{\alpha_1}(1 + rad(KG_0))$
elementary-p-Abelian
$Z_{\alpha_1-1}(G_1) = Z_{\alpha_1-1}(1 + rad(KG_0))$
... **stepwise elementary-p-Abelian**
$Z_2(G_1) = Z_2(1 + rad(KG_0))$
elementary-p-Abelian
$Z(G_1) = Z(1 + rad(KG_0))$

determines

$1 + rad(KZ(G_1))$

$\overline{K(G_1)}$
elementary-p-Abelian

1_{G_2}

determined by class number of G_1

$G_2 = 1 + rad(KG_1) = Z_{\alpha_2}(1 + rad(KG_1))$
elementary-p-Abelian
$Z_{\alpha_2-1}(G_2) = Z_{\alpha_2-1}(1 + rad(KG_1))$
... **stepwise elementary-p-Abelian**
$Z_2(G_2) = Z_2(1 + rad(KG_1))$
elementary-p-Abelian
$Z(G_2) = Z(1 + rad(KG_1))$

$1 + rad(KZ(G_2))$

$\overline{K(G_2)}$
elementary-p-Abelian

1_{G_3}

determined by class number of G_2

$\longrightarrow$ exp$(Z(\mathbf{s}_n))$ bounded
$\longrightarrow$ exp$(Z(G_n)) =$ exp$(Z(G_1))$

exponent and class of nilpotency unbounded

Matchmaking as a Combinatorial Exercise

I've browsed the combinations:
seven billion choose 2.
That's 24 quintillion –
it took some time to do.
The brute-force search was worth is, though,
to prove my theorem true:
"No couples in the space compares
to pairing me with you."

(unknown author, see [81])

6.4 Open topics and exercises

Let n be a natural number, p a prime number, G a finite p-group, K a field and $char(K) = p$.

Open-ended questions 6 *(i) Is the chain of solvable classes of $(G_n)_{n \in \mathbb{N}_0}$ strict monotone increasing? The author conjectures that the derived length is not bounded. Thus, – by a theorem of Gluck – the numbers $cd(G)$ of degrees of irreducible complex characters are also unbounded (because $2 \cdot cd(G) \geq st(G)$ is valid).*

(ii) Is the chain of Lie solvable classes of $((KG_n)^\circ)_{n \in \mathbb{N}_0}$ strict monotone increasing? The author conjectures that the Lie derived length is not bounded. As a consequence, the Wielandt length – as introduced by Barnes and Groves in [6] – associated to the chain is unbounded (because it leads to a bound for the derived length, see theorem 8.4 in the mentioned article).

(iii) For a fixed $r \in \mathbb{N}$ analyze whether the chain of exponents of the r-th derived subgroup of $(G_n)_{n \in \mathbb{N}_0}$ is strict monotone increasing.

(iv) For a fixed $r \in \mathbb{N}$ analyze whether the chain of exponents of the r-th term of the descending central chain of $(G_n)_{n \in \mathbb{N}_0}$ is strict monotone increasing.

(v) For every subgroup U of G focus on the iterative normalizers of U in $1 + rad(KG)$ (see exercise 229 and 230). If K is finite, then this chain reaches $1 + rad(KG)$ after finite many steps. How many steps are needed? Is it possible to describe the factor groups along this chain? What about the special case $U = G$?

(vi) Is the defect of subnormality of G in $1 + rad(KG)$ exactly $cl(1 + rad(KG))$? What is the defect for a (normal) subgroup of G in $1 + rad(KG)$?

A chain of p-groups 177

(vii) Determine the class number of $1 + rad(KG)$.

(viii) In [20] Coleman and Passman have proven that $Z_p \wr Z_p$ is involved in $1 + rad(KG) = G_1$. The author conjectures that he wreath product $\underbrace{C_p \wr ... \wr C_p}_{n-times}$ is involved in G_{n-1}. If the conjecture would be true, then some results within this chapter are proven differently: the chain of associated derived length, of nilpotency classes and of exponents of $(G_n)_{n \in \mathbb{N}_0}$ are not bounded (because all of these parameters are growing inside $\underbrace{C_p \wr \cdots \wr C_p}_{n-times}$ if n is growing). In addition, the complexity of the groups G_n would be presented: they would be linked to Sylow subgroups of the symmetric groups.

(ix) Let us focus again on the chain $(G_n)_{n \in \mathbb{N}_0}$ and here on the exponents alongside the lower central series of G_n. As proven for the upper central chain, is it true that the exponents of derived factor group related to the chain are bounded and that the factors groups beneath this factor group are elementary-p-Abelian?

(x) What can be said about the dimensions of the irreducible complex characters related to the chain $(G_n)_{n \in \mathbb{N}_0}$? Is the sum of all dimensions resp. the maximal dimension of the irreducible characters increasing? What about their average dimension?

(xi) Let p be a prime number, q a power of p and P be a finite p-group. P is isomorphic to a subgroup of p-Sylow subgroup of the symmetric group $S_{|P|}$ (Theorem of Cayley), the general linear group $GL(|P|, GF(q))$ and the unit group $U(GF(q)P)$ of the modular group algebra $GF(q)P$ (based on a theorem of Wallace). Do some more non-trivial series of p-Sylow subgroups exist which let P be embedded as a subgroup, too? All of them are grandfathers of all p-groups. For the mentioned ones we have proven that the exponents of the ideal of class sums is exactly p. Is this true for all such grandfather series? What is the definition of a grandfather series?

(xii) Let R be a radical algebra. Is there a connection between the Baer-condition of the Lie algebra R° and the adjoint group $R^\star$?

(xiii) Let $r \in \mathbb{N}$. Within remark 6.3.4 we have bounded the sequence of exponents of the r-th centers related to $(G_n)_{n \in \mathbb{N}_0}$. Is it true that the sequence is convergent? If this is the case, then determine its limit.

Excercise 212 For a fixed $r \in \mathbb{N}$ prove that the chain of r-th terms of the descending central chain related to $(G_n)_{n \in \mathbb{N}_0}$ is strict monotone increasing. (Hint: Prove that otherwise the class of nilpotency is bounded.) Is it possible to transfer the argumentation to $((KG_n)^\circ)_{n \in \mathbb{N}_0}$?

Excercise 213 Let us focus on the the chain $(G_n)_{n\in\mathbb{N}_0}$ and let $r \in \mathbb{N}$. Prove the following fact: If the chain of r-th derived subgroups related to $(G_n)_{n\in\mathbb{N}_0}$ is bounded, than the derived length related to $(G_n)_{n\in\mathbb{N}_0}$ is bounded, too. Is it possible to transfer the argumentation to $((KG_n)^\circ)_{n\in\mathbb{N}_0}$?

Excercise 214 Let us focus on the irreducible complex characters related to the chain $(G_n)_{n\in\mathbb{N}_0}$. Prove that the number of all resp. of all linear ones is strictly increasing. (Hint: part (xi) of proposition 6.1.2 for the class numbers, its proof for using that $(G_1)' \cdot G$ is not identical to G_1 which can be used to prove that $\mid G/G' \mid$ is smaller than $\mid G_1/(G_1)' \mid$)

Excercise 215 (submitted by Salvatore Siciliano on researchgate) Take a non-nilpotent Lie algebra L satisfying an Engel condition and, for every $n > 0$, consider the quotient of L with respect to the n-th term of the descending central series. For each n obtain a nilpotent Lie algebra of class n satisfying the Engel condition you started with. Now, by using suitable direct sums construct an example such that $(L_n)_{n\in\mathbb{N}}$ is a monotone increasing sequence of finite-dimensional nilpotent Lie algebras over a field of characteristic $p > 0$ such that the Engel length of the sequence is bounded but the class of nilpotency is unbounded.

Excercise 216 Study the article [17] of Camina concerning the Wielandt length of finite groups. What is the definition of the Wielandt length? Use lemma 1 within the article to prove the following statement: The Wielandt length associated to the chain $(G_n)_{n\in\mathbb{N}_0}$ is unbounded. (Hint: The Wielandt length is a bound for the Baer-length. Use exercise 218.)

Excercise 217 Let us again focus on the breadth associated to the chain $(G_n)_{n\in\mathbb{N}_0}$. Do a research in the literature to find theorems of Vaughan-Lee, Cartwright, Neumann, Wiegold and Shepperd to bound the order of the derived subgroup of G in terms of the breadth. Use the results for the derived subgroups related to the chain $(G_n)_{n\in\mathbb{N}_0}$ to prove that the breadth related to the chain $(G_n)_{n\in\mathbb{N}_0}$ are unbounded.

Excercise 218 Prove the statements mentioned before proposition 6.1.2 concerning Baer-topics of Lie algebras. How is the class of nilpotency of $(KG)^\circ$ resp. the class of nilpotency of $E(KG)$ linked to n-Baer Lie group algebras $(KG)^\circ$? (Hint: use [54]) Transfer the definition of the Baer condition and its equivalent definition based on n-stable series of $ad(x)$ to groups by using group commutators instead of Lie commutators. The basic definition should be that every cyclic subgroup is subnormal of defect n. Prove that a n-Baer group is n-Engel, and that a group of Engel-length n and derived length d is $(d \cdot n)$-Baer. In addition, deduce that the Baer-length associated to the chain $(G_n)_{n\in\mathbb{N}_0}$ is unbounded.

A chain of p-groups   179

Excercise 219 *Use exercise 218 to prove the following theorem: There is a function $h : \mathbb{N} \longrightarrow \mathbb{N}$ such that for all $n \in \mathbb{N}$ the following statement is true: If $(KG)^{\circ}$ is n-Baer, then $E(KG)$ is $h(n)$-Baer.*

Excercise 220 *Prove or disprove the following argumentation: This exercise is dedicated to the converse of exercise 219. Let $E(KG)$ or equivalent $1 + rad(KG)$ be n-Baer.*

At first we want to show that there is a function $g : \mathbb{N} \longrightarrow \mathbb{N}$ such that for all $n \in \mathbb{N}$ the following statement is true: If U is a subgroup of $1+rad(KG)$, then the defect of subnormality of U is bounded by $g(n)$. We argue by induction on $|U|$. Take a maximal normal subgroup N of U which is of index p. Let $u \in U$ such that $U = N \cdot \langle u \rangle_{\mathcal{G}}$ is true. We use lemma 2.2. in [55] to deduce that the defect of subnormality of U – which is the join of N and $\langle u \rangle_{\mathcal{G}}$ and in which N is normal – is bounded by the defect of N multiplied by the defect of $\langle u \rangle_{\mathcal{G}}$. The first factor can be treated by induction and the second factor is less than n by our assumption.

We continue by using the well-known result of Roseblade (see [58]). Thus, the nilpotency class of $1 + rad(KG)$ can be bounded by $h(g(n))$ for another function $h : \mathbb{N} \longrightarrow \mathbb{N}$.

Now we use the theorem of Du (see [24]) to deduce that the class of nilpotency of $(KG)^{\circ}$ is bounded by $h(g(n))$. Thus, $(KG)^{\circ}$ is $h(g(n))$-Baer.

Excercise 221 *Prove that for a nilpotent Lie algebra L all its subalgebras are subnormal of defect/index bounded by the class of nilpotency of L. Is this defect/index associated to the chain $((KG_n)^{\circ})_{n \in \mathbb{N}}$ bounded or unbounded? Transfer the results to $(G_n)_{n \in \mathbb{N}}$.*

Excercise 222 *Investigate whether the inner, outer and full automorphism groups related to the chain $(G_n)_{n \in \mathbb{N}_0}$ are (strict) monotone increasing. Exercise 43 might be of help.*

Excercise 223 *Prove or disprove: G and $1 + rad(KG)$ are isoclinic.*

Excercise 224 *Prove or disprove: G and $1 + rad(KG)$ are isomorphic.*

Excercise 225 *Let G be cyclic. What is the probability that an element generates G? What is the probability that two elements of G are of the same order? What is the probability that for two elements g, h of G the condition $\langle g \rangle_{\mathcal{G}} = \langle h \rangle_{\mathcal{G}}$, $\langle g \rangle_{\mathcal{G}} <> \langle h \rangle_{\mathcal{G}}$, $\langle g \rangle_{\mathcal{G}} < \langle h \rangle_{\mathcal{G}}$, $\langle g \rangle_{\mathcal{G}} \leq \langle h \rangle_{\mathcal{G}}$, $\langle g \rangle_{\mathcal{G}} > \langle h \rangle_{\mathcal{G}}$ resp. $\langle g \rangle_{\mathcal{G}} \geq \langle h \rangle_{\mathcal{G}}$ is valid? Analyze these probabilities if p or G is getting large.*

Excercise 226 *Let G be Abelian. What is the probability that an element $g \in G$ is chosen such that $o(g) = exp(G)$ is true? Analyze this probability if p or G is getting large.*

Excercise 227 Let K be finite. This exercise is dedicated to the degree of commutativity of G denoted by $d(G)$. Prove/disprove and solve the following facts/statements (and use the master thesis [18] as a basis):

(i) $d(Q_8) = \frac{5}{8}$

(ii) $d(A_4) = 1$

(iii) $d(E(KG)) = d(1 + rad(KG)) = d(rad(KG)^\star)$

(iv) For every subgroup H of G the statement $d(G) \leq d(H)$ is valid.

(v) $d(G) = \frac{c(G)}{|G|}$

(vi) For every subgroup H of G the statement $c(G) \leq c(H) \cdot \frac{|G|}{|H|}$ is valid. (One aspect of the open topic (vii).)

(vii) Use the previous item to bound $c(1 + rad(KG))$.

(viii) $d(G) \leq \frac{3}{2 \cdot p^{b(G)}}$

(ix) $d(G) = 1$ iff G is nilpotent.

(x) A group G exists such that $d(G) = \frac{9}{10}$ is valid.

Excercise 228 Use exercise 179 to describe the behavior of the chain of nilpotency classes resp. the maximum of all nilpotency classes of the elements of $(ZG_n)_{n \in \mathbb{N}_{\geq 2}}$. For the analysis of nilpotency classes consider direct products and the formula mentioned within exercise 179. Compute the class explicitly using the results for $Z(G_1)$ and $Z(G_n)$ for $n \geq 2$. For the second topic use the connection to the exponent.

Excercise 229 This exercise is dedicated to item (viii) of the open questions mentioned in this chapter. We want to prove that the chain $(G \cdot Z_n(1 + rad(KG)))_{n \in \mathbb{N}}$ is of length $cl(1 + rad(KG))$ if the field K has more elements than the group G. Let us assume that an element $n < cl(1 + rad(KG)) := c$ exists such that $G \cdot Z_n(1 + rad(KG)) = 1 + rad(KG)$. Then, $G \cdot Z_{c-1}(1 + rad(KG)) = 1 + rad(KG) = Z_c(1 + rad(KG))$ is valid. Based on the theorem of [24] we know that $Z_{c-1}(1 + rad(KG)) = 1 + Z_{c-1}(rad(KG)^\star) = 1 + Z_{c-1}(rad(KG)^\circ)$ is true. The dimension of $Z_c(1 + rad(KG))/Z_{c-1}(1 + rad(KG))$ is at least 1. We conclude $|G/(G \cap Z_{c-1}(1 + rad(KG)))| = |Z_{c-1}(1 + rad(KG))/Z_c(1 + rad(KG))| \geq |K|$. Prove this argumentation in details!

The idea for proving this statement for arbitrary fields is to tensor the group algebra KG by a suitable field extension L which is large enough. The upper central chain of $(KL)^\circ = ((KG) \otimes L)^\circ$ is $Z_n(rad(KG)^\circ) \otimes L =$

$Z_{c-1}(rad(KL)^\circ)$. The class of nilpotency does not change. If we could prove that $G \cdot Z_n(1 + rad(KG)) = 1 + rad(KG)$ leads to $G \cdot Z_n(1 + rad(KL)) = 1 + rad(KL)$, then we would have proven this statements for arbitrary fields. Prove the argumentations in details if possible.

Excercise 230 This exercise is also dedicated to item (viii) of the open questions mentioned in this chapter. We present an idea for determining the iterative chain of normalizers of G in $1 + rad(KG)$.

We begin with a definition: Let G be a group, U a subgroup of G and $n \in \mathbb{N}_{\geq 2}$. We define $N_G^{(1)}(U) := N_G(U)$ and $N_G^{(n)}(U) := N_G^{(n-1)}(U)$. We call this chain the chain of iterated normalizers of U in G. In addition, let $Z_{n+1}^G(1 + rad(KG)) := \{x \mid \forall g \in G : [x,g] \in Z_n^G(1 + rad(KG))\}$ for all $n \in \mathbb{N}$ which is normal in $G \cdot Z_{n+1}^G(1 + rad(KG))$ for all $n \in \mathbb{N}$.

The following lemma is a fundamental tool: Let p be a prime number, G a finite p-group, K a finite field and $char(K) = p$, $n \leq cl(1 + rad(KG)) - 1$ and $a \in N_{1+rad(KG)}(G \cdot Z_n^G(1 + rad(KG)))$. We define the action

$$\delta : G \longrightarrow (G \cdot Z_n^G(1 + rad(KG)))/Z_n^G(1 + rad(KG)) \text{ by}$$

$$g\delta : hZ_n^G(1 + rad(KG)) \mapsto (g^{-1}hg^a)Z_n^G(1 + rad(KG)).$$

The following statements are equivalent:

(i) δ possesses a fixed point.

(ii) $\hat{\delta}$ possesses a fixed point.

(iii) $a \in G \cdot Z_{n+1}^G(1 + rad(KG))$.

Proof. Let us remark as long as $n \leq cl(G)$ then $(G \cdot Z_n^G(1+rad(KG)))/Z_n^G(1+rad(KG))$ is isomorphic to a proper factor group of G which is $G/(G \cap Z_n^G(1 + rad(KG)))$.

The statements (i) and (ii) are equivalent based on theorem 1.3.5. Let us focus on the equivalence of (ii) and (iii). The following relations are true:

δ possesses a fixed point iff
$\exists h \in G \forall g \in G\, g^{-1}hg^a Z_n^G(1 + rad(KG)) = hZ_n^G(1 + rad(KG))$ iff
$\exists h \in G \forall g \in G\, (ha^{-1})^g Z_n^G(1 + rad(KG)) = (ha^{-1})Z_n^G(1 + rad(KG))$ iff
$\exists h \in G \forall g \in G\, [ha^{-1}, g] \in Z_n^G(1 + rad(KG))$ iff
$\exists h \in G\, ha^{-1} \in Z_{n+1}^G(1 + rad(KG))$ iff
$a^{-1} \in G \cdot Z_{n+1}^G(1 + rad(KG)) = Z_{n+1}^G(1 + rad(KG)) \cdot G$ iff
$a \in G \cdot Z_{n+1}^G(1 + rad(KG)) = Z_{n+1}^G(1 + rad(KG)) \cdot G$.

The idea is to use this lemma as follows:
Let $a \in N_{1+rad(KG)}(G \cdot Z_n^G(1 + rad(KG)))$ like $a = \sum_{g \in G} k_x x$. We use the action $g\delta : h Z_n^G(1 + rad(KG)) \mapsto (g^{-1} h g^a) Z_n^G(1 + rad(KG))$. The equation $g^{-1} a g^a = a$ is valid. Thus, $g^{-1} \sum_{g \in G} k_x x g^a = \sum_{g \in G} k_x x$ is true. We deduce $\sum_{g \in G} k_x g^{-1} x g^a = \sum_{g \in G} k_x x$. Thus, $\sum_{g \in G} k_x x \cdot Z_n^G(1 + rad(KG)) = \sum_{g \in G} k_x (g^{-1} x g^a) Z_n^G(1 + rad(KG)) = g^{-1} Z_n^G(1 + rad(KG))(\sum_{g \in G} k_x x Z_n^G(1 + rad(KG)))(g^a Z_n^G(1 + rad(KG)))$ is true.

Excercise 231 *This example is based on a communication with Stewart Stonehewer for which I want to say thank you now. Let H be a cyclic group of order 2 and let K be the dihedral group of order 8 with y a non-central element of order 2. Then let G be the wreath product of H by K with base group B. G is of order 2048. Then $J := \langle H, Hy \rangle_{\mathfrak{G}}$ is normal in B which is normal in G. But the normalizer of J in G is $B \cdot \langle y \rangle_{\mathfrak{G}}$ which is not normal in G. Thus, the defect of subnormality is 2 but the chain of iterated normalizers has length ≥ 3. Prove all of these statements in details. Is it possible to extend this example to arbitrary prime numbers?*

Excercise 232 *Prove remark 6.2.13 in details.*

Excercise 233 *Prove corollary 6.2.12 in details.*

Excercise 234 *Let a, n and b_i for $i \in \underline{n}$ some natural numbers. Based on the equation $ab_1 - 1 = (a - 1)b_1 + (b_1 - 1)$ rewrite $ab_1 \ldots b_n - 1$. In what way is this exercise relevant within this chapter?*

Excercise 235 *Let n and a_i, b_i for $i \in \underline{n}$ some natural numbers. Based on the equation $a_2 b_2 - a_1 b_1 = (a_2 - a_1) b_2 + a_1 (b_2 - b_1)$ rewrite $a_n b_n \ldots a_1 b_1$. In what way is this exercise relevant within this chapter?*

Excercise 236 *This exercise is related to the parts (i) and (ii) of the open questions and exercises of this chapter: Are the chains of solvable classes of $(G_n)_{n \in \mathbb{N}_0}$ resp. $((KG_n)^\circ)_{n \in \mathbb{N}_0}$ strict monotone increasing? The idea is to prove that $st(G) < st(1 + rad(KG))$ and $st(G) < st((KG)^\circ)$ are valid. From both inequalities the original topic can be proven.*

Let us start with $st(G) < st((KG)^\circ)$. If G' is Abelian, then $st(G) = 2$ is valid. Within [65] it is proven that $st((KG)^\circ)$ is not smaller than the upper integral part of $\log_2(p + 1)$. For $p > 3$ we can assume that G' is not Abelian. In [69] it is proven that G' is bounded if this lower bound is met. This is not possible on long term because of part (iii) of proposition 6.1.2. If G' is not Abelian, then we can use an induction argument to show $st(G') < st((KG')^\circ)$. In [65] it is proven for $p > 2$ that

A chain of p-groups 183

$rad(KG')KG + Z(KG) = (KG \circ KG) + Z(KG)$ is valid. From this the first inequality can be deduced for $p > 2$.

In [70] a lower bound for $st(E(KG))$ is presented similar to the one for $st((KG)^\circ)$ and also for which groups G this bound is met. An analogue argumentation would be valid if the condition $rad(KG')KG + Z(KG) = (KG \circ KG) + Z(KG)$ could be transferred to the group of units.

Excercise 237 *Prove that the normal closure of G in $1 + rad(KG)$ is smaller than $1 + rad(KG)$.*

Excercise 238 *Proof definition and remark 6.2.5 in details.*

Excercise 239 *Proof proposition 6.2.10 in details.*

Excercise 240 *Prove part (iv) of proposition 6.1.2 in details.*

Excercise 241 *Do a research in the literature and find a connection between the nilpotency class and breadth of a finite group. Apply this connection to proposition 6.1.2.*

List of Figures

local group algebras 17
group of units and the star group 20
cores and normalizers 26
cores and normalizers within $E(GF(2)Q_8)$ 28

construction of an end-commutable ordering for a conjugacy class
  of D_{16} 46
a special chain of subnormal subgroups 48
orders of conjugacy class sums and centralizers 56

maximal possible exponents 70
elementary-Abelian centers 71
main results on exponents 95

an ideal within the center of the radical 101
structure of the center 107
the class-graph 113
Isoclinism 119

elementary-Abelian factors alongside the upper central chain ... 139

properties of the chain of iterated p-groups 174
behaviour of the center of the chain $(G_n)_{n \in \mathbb{N}_0}$ 176

Bibliography

[1] Angela Albrecht, Die Struktur der Einheitengruppe endlicher kommutativer Gruppenringe, Diplomarbeit, Kiel, 1988

[2] Bernhard Amberg, Yaroslav Sysak, Associative rings with metabelian adjoint group, Journal of Algebra, Vol. 277, Issue 2, 456-473, 2004

[3] Bernhard Amberg, Yaroslav Sysak, Radical Rings with Engel Conditions, Journal of Algebra, No. 231, 364-373, 2000

[4] Bernhard Amberg, Yaroslav Sysak, Radical Rings with Soluble Adjoint Groups, Journal of Algebra, No. 247, 692-702, 2002

[5] Czeslaw Baginski, Janos Kurdics, On the center of the modular group algebra of a finite p-group, Journal of Algebra and its applications, Vol. 13, No. 04, 2014

[6] Donald W. Barnes, Daniel Groves, The Wielandt subalgebra of a Lie algebra, J. Aust. Math. Soc., No. 74 313-330, 2003

[7] Berkovich, Yakov, Groups of prime power order, vol. 1, de Gruyter Expositions in Mathematics, 46, Walter de Gruyter, Berlin, 2008

[8] J. C. Bioch, On n-isoclinic groups, Journal reine and angewandte Mathematik, 182, 130-141, 1940

[9] Adalbert A. Bovdi, Zoltan Patay, On the central units of a modular group algebra, Acta. Sci. Math. (Szeged) 63, 71-82, 1997

[10] Adalbert A. Bovdi, Zoltan Patay, Ulm-Kaplansky invariant of the center of the group of units of modular group ring, Dep UkrNIINTI 360, Uk-85, 1-35, 1996

[11] Adalbert A. Bovdi, Piroska Lakatos, On the exponent of the group of normalized units of a modular group algebra, Publicationes mathematicae 52(3-4), 409-411, January 1993

[12] Adalbert A. Bovdi, Zoltan Patay, The structure of the center of the multiplicative group of the group ring of a p-group over a ring of characteristic p, Vestsi Akad. Nauk BSSR Ser. Fiz.-Mat. Nauk 1, 1978, 5-11

[13] Adalbert A. Bovdi, A. Szakacs, A basis for the unitary subgroup of the group of units in a finite commutative ring, Publ. Math. Debrecen 46 (1-2), 97-120, 1995

[14] Adalbert A. Bovdi, César Polcino Milies, Normal subgroups of the group of units in group rings of torsion groups, Publicationes mathematicae 59 (1), August 2001

[15] Adalbert A. Bovdi, J. Kurdics, Lie properties of the group algebra and the nilpotency class of the group of units, Journal of Algebra 212, 28-64, 1999

[16] V. Bovdi, A. L. Rosa, On the oder of the unitary subgroup of modular group algebra, arxiv.org, Cornell University Library, 09/2000

[17] A. R. Camina, The Wielandt Length of Finite Groups, Journal of Algebra, No. 15, 142-148, 1970

[18] Anna Castelaz, Commutativity degree of finite groups, Master of arts, Wake Forest University, 2010

[19] Donald B. Coleman, On the modular group ring of a p-group, Proc. Amer. Math. Soc. 15, 511-514, 1964

[20] Donald B. Coleman, D.S. Passman, Units in modular group rings, Proc. Amer. Math. Soc. 25, No.3, 510-512, 1970

[21] Donald B. Coleman, Robert Sandling, Mod 2 group algebras with metabelian unit groups, Journal of Pure and Applied Algebra 131, no. 1, 25-36, 1998

[22] W. E. Deskins, Finite Abelian groups with isomorphic group algebras, Duke Math. J., Volume 23, Number 1, 35-40, 1956

[23] L.E. Dickson, Modular theory of group matrices, Trans. Amer. Math. Soc. 8, 389-398, 1907

[24] Xiankun Du, The centers of a radical ring, Canad. Math. Bull. 35, no. 2, 174-179, 1992

[25] Philip Hall, The classification of prime-power groups, Journal für die reine und angewandte Mathematik, 182, 130-141, 1940

[26] Bertram Huppert, Endliche Gruppen I, Springer-Verlag, Berlin, 1967

[27] I. M. Isaacs, Groups with many equal classes, Duke Math. J., 37(3), 501-506, 1970

[28] Kenta Ishikawa, Finite p-groups up to isoclinism which have only two conjugacy lengths, Journal of algebra 220, 333-345, 1999.

[29] Zvonimir Janko, Classification of finite p-groups with cyclic intersection of any two distinct conjugate subgroups, Glasnik Matematicki, Vol. 50(70), 101-161, 2015

[30] S. Jennings, The structure of the group ring of a p-group over a modular field, Trans. Amer. Math. Soc. 50, 175-185, 1941

[31] S.A. Jennings, Central chains of ideals in an associative ring, Duke Math. Journal 9, 1942 341-355

[32] S.A. Jennings, Radical rings with nilpotent associated group. Trans. Roy. Soc. Can. ser. III 49, 1955, 31-38

[33] D. L. Johnson, The modular group-ring of a finite p-group, Proceedings of the AMS, volume 68, number 1, January 1978

[34] Tibor Juhász, Derived lengths of symmetric and skew symmetric elements in group algebras, leroy.perso.math.cnrs.fr, Talks, Juhasz

[35] Gregory Karpilovsky, The Jacobson Radical of Group Algebras, Elsevier, 1987

[36] Hans-Georg Knoche, Über den Frobenius'schen Klassenbegriff in nilpotenten Gruppen, Mathematische Zeitschrift, Band 55, Heft 1, 71-83, 1952

[37] Robert L. Kruse, David T. Price, Nilpotent rings, Gordon and Breach, Science Publishers, Inc., New York, 1969

[38] J. Kurdics, Engel properties of group algebras II, Journal of Pure and Applied Algebra, No. 133, 1998, 179-196

[39] H. Laue, Assoziative Algebren, Vorlesung am Mathematischen Seminar der CAU zu Kiel, WS 2010/2011

[40] H. Laue, On the associated Lie ring and the adjoint group of a radical ring, Canadian mathematical bulletin, Vol. 27, No. 4, 1984, 217 ff.

[41] H. Laue, Lie-Algebren, Vorlesung am Mathematischen Seminar der CAU zu Kiel, WS 2008/2009

[42] M. R. Vaughan-Lee, Breadth and commutator subgroups of p-groups, Journal of algebra 32, 278-285, 1974

[43] M. R. Vaughan-Lee, J. Wiegold, Breadth, class and commutator subgroups of p-groups, Journal of algebra 32, 268-277, 1974

[44] F. Levin, G. Rosenberger, Lie metabelian group rings, Group and semigroup rings, North-Holland, 153-161, 1986

[45] Mark. L. Lewis, Semi-extraspecial groups, Cornell University, https://arxiv.org/abs/1709.03857, 2017

[46] Mark L. Lewis, Semi-extraspecial p-groups, Kent State University January, 4th Biennial International Group Theory Conference 2017 (4BIGTC2017) - Kuala Lumpur, Malaysia, 2017

[47] Mark L. Lewis, Generalizing Camina groups and their character tables, J. Group Theory 12, 209-218, 2009

[48] Mark L. Lewis, Centralizers of Camina p-groups of nilpotence class 3, Journal of Group Theory, volume 21, issue 2, pages 319 ff., 2015

[49] L. E. Moran, The modular group ring of a p-group, M. Phil. Thesis, University of Nottingham, 1972

[50] Yasushi Ninomiya, Nilpotency indicies of the radicals of finite p-solvable group algebras, I, J. Austral. Math. Soc. 71, 117-133, 2001

[51] Donald S. Passman, The algebraic structure of group rings, Wiley-Interscience Publication, New York, 1977

[52] Donald S. Passman, Group rings satisfying a polynomial identity, Journal of algebra, N0. 20, 103-117, 1972

[53] K.R. Pearson, On the units of a modular group ring II, Bull. Austral. Math. Soc. 8, 435-442, 1973

[54] Eliyahu Rips, Aner Shalev, The Baer condition for Group Algebras, Journal of Algebra, No. 140, 1991

[55] Derek S. Robinson, Joins of subnormal subgroups, Illinois J. Math., Volume 9, Issue 1, 144-168, 1965

[56] Joseph J. Rotman, An introduction to the theory of groups, Springer Verlag, New York, 1995

[57] Robert Sandling, Units in the modular group algebra of a finite abelian p-group, J. Pure Appl. Algebra 33, 337-346, 1984

[58] J. E. Roseblade, On groups in which every subgroup is subnormal, Journal of Algebra, Volume 2, Issue 4, 402-412, 1965

[59] Scholz, Karsten, Zentralreihen in Radikalringen, Diplomarbeit, Mathematisches Seminar der CAU zu Kiel, 1996

[60] Aner Shalev, Meta-abelian unit groups of group algebras are usually abelian, Journal of Pure and Applied Algebra 72, 295-302, 1991

[61] Aner Shalev, Applications of dimension and Lie dimension subgroups to modular group algebras, Proceedings of the Amitsur conference in ring theory, Jerusalem, 85-94, 1989

[62] Aner Shalev, Avinoam Mann, The nilpotency class of the unit group of a modular group algebra II, Israel J. Math., no. 3, 67-77, 1990

[63] Aner Shalev, The nilpotency class of the unit group of a modular group algebra III, Arch. Mat. Vol. 60, 136-145, 1993

[64] Aner Shalev, Dimension Subgroups, Nilpotency Indices, and the Number of Generators of Ideals in p-Group Algebras, Journal of Algebra 129, 412-438, 1990

[65] Aner Shalev, The derived length of Lie soluble group rings I, Journal of Pure and Applied Algebra 78, 291-300, 1992

[66] Aner Shalev, Lie dimension subgroups, Lie nilpotency indices, and the exponent of the group of normalized units, Journal of London Mathematical Society, No. 2, 23-36, 1991

[67] Dane Christian Skabelund, Character Tables of Metacyclic Groups, Master of Science, Brigham Young University - Provo, 2013

[68] Francesco Catino, Ernesto Spinelli, A Note on Strong Lie Derived Length of Group Algebras, Bollettino dell'Unione Matematica Italiana, Serie 8, Vol. 10-B, n.1, p. 83-86, Unione Matematica Italiana, 2007

[69] Francesco Catino, Ernesto Spinelli, On the derived length of the unit group of a group algebra, Journal of Group Theory 13, 577-588, 2010

[70] Ernesto Spinelli, Group algebras with minimal Lie derived length, Journal of Algebra 320, 1908-1913, 2008

[71] Bernd Stellmacher, Hans Kurzweil, Theorie der endlichen Gruppen, Springer-Verlag, Berlin, Heidelberg, 1998

[72] Mathias Theede, Die aufsteigende Zentralreihe in Einheitengruppen modularer Gruppenalgebren fur Klassen metabelscher p-Gruppen, Dissertation, Kiel, 2016

[73] Juhász Tibor, On the derived length of Lie solvable group algebras, dissertation, Debrecen, 2006

[74] D.A.R. Wallace, On the radical of a group algebra, Proc. Amer. Math. Soc. 12, 133-137, 1961

[75] A. J. Weir, Sylow p-subgroups of the classical groups over finite fields with characteristic prime to p, Proc. Amer. Math. Soc. vol.6, 529-533, 1955

[76] Sven Wirsing, Über die Struktur der Solomon-Tits-Algebren der symmetrischen Gruppen, disserta, 2015

[77] https://mathoverflow.net/questions/122553/p-group-with-large-center

[78] https://math.stackexchange.com/questions/3190193/centralizers-in-p-groups

[79] https://math.stackexchange.com/questions/3244149/two-p-groups-of-exponent-p-with-same-number-of-conjugacy-classes-but-non-isomo

[80] http://www.copsmodels.com/kenpearson.htm

[81] https://mathwithbaddrawings.com/2018/02/12/love-poems-for-mathematicians/

Index

associated Lie algebra
    class of nilpotency, 165
    definition, 165
    lower central chain, 165
    nilpotent, 165
    upper central chain, 165
associative algebra
    Jacobson radical, 165
    nil, 165
    nilradical, 165
    radical algebra, 165
augmentation, 14
augmentation ideal, 14

center of $E(KG)$, 27
center of the radical
    reduction of the determination of the exponent, 54
    torsion group, 53
        exponent, 55
centralizers in $E(KG)$, 29
chain of p-groups
    basic facts, 160
    class sums, 165
    definitions, 159
circle group
    Du, 167
    Jennings, 166
    Laue, 167
    main theorem, 167
    operation on the additive group, 166
class number, 27
commutator, 27
conjugacy classes and commuting elements, 50

conjugates in $E(KG)$, 30
cores, 20

D.B. Coleman, 24
decomposition of the center
    a direct decomposition, 102
    Abelian groups, 104
    class-graph, 113
    complements, 106
    dihedral groups, 116
    examples for the maximal case, 115
    examples for the minimal case, 115
    invariants for Abelian group algebras, 104
    invariants of the center, I, 109
    invariants of the center, II, 112
    monotony, 108
    quaternion groups, 116
    semi-dihedral groups, 116
Du, Xiankun, 166, 167

end-commutable ordering
    commuting elements, 50
    compatabilities, 51
    conjugacy classes, 45
    construction method, 45
    group algebra, 38
    maximal, 51
    nilpotency criterion, 49
    nilpotent group, 45
    operation of $Aut(G)$, 40
    p-powers, 38
    polycyclic group, 51
end-commutable orderings

definition, 37
enhanced group action, 24
enhancement of conjugacy classes of G in $E(KG)$, part I, 32
enhancement of conjugacy classes of G in $E(KG)$, part II, 32
exponent of the center of the racikal bounds, 57
exponent of the center of the radical
  (generalized) Camina groups, 121
  bounds, 88
  dihedral groups, 67
  every Abelian characteristic subgroup is cyclic, 79
  every Abelian normal subgroup can be $\mathcal{G}$-generated by 1 or 2 elements, 80
  extra-special p-groups, 79
  general wreath products, 85
  group extensions, 89
  Hamiltonian p-groups, 79
  identical to exponent of $Z(G)$, 73
  isoclinic groups, 119
  isoclinism, 119
  linear group, 72
  linear, symplectic, orthogonal and unitary groups, 87
  metacyclic groups, 80
  quaternion groups, 67
  regular p-group, 74
  regular wreath products, 86
  semi-dihedral groups, 67
  semi-extra-special groups, 120
  special p-group, 73
  symmetric groups, 87
  the maximal exponent, 67
  the minimal exponent, 70
  ultra-special groups, 120
  VZ-groups, 121
  wreath product and conjugation, 87
  wreath products and faithful operation, 86
  wreath products and operation on cosets, 86
exponent the center of the radical central products, 76

fixed point lemma, 24

group
  breadth, 159
  degree of commutativity, 160
  derived length, 160
group of units
  class of nilpotency II, 156
  unitary subgroup, 157

idempotent elements and subgroups, 13

Jacobson, Nathan, 165, 166
Jennings, Stephen Arthur, 166, 167

Köthe, Gottfried, 165

L.E. Dickson, 14
Laue, Hartmut, 166, 167
lemma of Wallace, 17
Lie algebra
  Baer condition, 160
local group algebra, 18

nilpotence criteria by conjugation, 49
normal and central in $E(KG)$, 20
normal subgroups and ideals, 14
normalizers in $E(KG)$, 24

quasi regular, 11
quasi regular group
  commutator, 166
  conjugation, 166
  definition, 166
  inverse, 166
  lower central chain, 166
  nilpotency class, 166
  upper central chain, 166

Scholz, Karsten, 167

semidirect decomposition, 12
star composition, 11
star group, 11
star regular, 11

theorem of Pearson, 22

unit groups
    class of nilpotency, 137
    cyclic p-power subgroup, 142
    cyclic derived subgroup, 134
    cyclic Frattini subgroup, 135
    exponent, 143
    exponents alongside the central chain, 139
    meta-Abelian, 135
    meta-cyclic, 135
    special and extra-special, 144
    wreath product, 134

van der Waerden, Bartel Leendert, 166